# PLANNING IN CHINESE AGRICULTURE

# PLANNING IN
# CHINESE AGRICULTURE

*Socialisation and the Private Sector,*
*1956 - 1962*

## KENNETH R. WALKER

Lecturer in Economics, School of Oriental and African Studies
University of London

FRANK CASS & COMPANY LTD.
1967

First published in 1965 by
Frank Cass & Co. Ltd., 10 Woburn Walk,
London, W.C.1

*This essay is dedicated
to my parents*

# CONTENTS

# TABLES

# Preface

In 1959, when the School of Oriental and African Studies in the University of London decided to establish a group of economists to examine the modern problems of Asian countries, I was fortunate enough to be appointed and chose to study the Chinese economy. This short monograph is the first result of that decision. It would not have been written without the encouragement and advice of too many people to name here. In the first two years Dr. J. D. Chinnery and Mrs. Y. C. Liu in London, and later Mr. Y. C. Yin in Harvard, devoted hours of their valuable time to teach me to read modern Chinese economic source materials. To them, above all, I now extend my grateful thanks. Since the formation of the Department of Economic and Political Studies at the School I have been constantly supported by my colleagues, benefitting greatly from their criticisms and suggestions made both privately and in seminars. In this respect I must record a special debt to my colleague and friend Mr. T. J. Byres for the stimulation of innumerable discussions. Mr. J. Lust, head of the Chinese section of the Library at the School, gave his time generously to obtain Chinese materials needed to carry out the research. Mrs. K. Austin prepared the manuscript for publication, skilfully coping with handwriting which has come to resemble Chinese characters. None of these people must in any way be associated with the views and conclusions expressed in this work.

Kenneth R. Walker
School of Oriental and African Studies,
University of London.
*November*, 1963.

# Introduction

During the short period between the establishment of the Chinese Communist Government in 1949 and Autumn 1958 the economic organisation of agriculture experienced a spectacular revolution. The tiny, fragmented peasant farms of 1·34 hectares,[1] privately owned either by landlords or peasant occupiers, were replaced in turn by mutual aid teams, agricultural producers' co-operatives, collective farms and, finally, by the people's communes. Each stage brought larger agricultural planning units and more socialisation of the factors of production. The communes were to be the final stage in the full socialisation of agriculture. Already in Autumn 1958 party cadres were acclaiming the advent of the stage of communism. Though this was premature, at east the sprouts of communism had appeared. The Governmlent viewed the new form of agricultural organisation as the key to controlling the fickle environment and to supplying abundant food, both for consumption and development. In Marxist terms, productive relations no longer fettered the productive forces of China.

This relationship between socialisation and agricultural development is one of the most crucial (and politically sensitive) problems discussed by those who seek solutions to promote economic development in poor, agrarian economies. Most economists[2] have stressed that to secure economic progress in agriculture there must be fundamental changes in the institu-

[1] This was the medium size of farm in J. L. Buck's survey of 16,786 farms in 22 Provinces of China, 1929–33. J. L. Buck, *Land Utilisation in China*, Shanghai, 1937.

[2] For example, N. Georgescu-Roegen, *Economic Theory and Agrarian Economics*. Oxford Economic Papers 1960, No. 1, pp. 1–40. T. Balogh, *Agricultural and Economic Development: Linked Public Works*, ibid., 1961, No. 1, pp. 27–42. V. M. Dandekar, *Economic Theory and Agrarian Reform*, ibid., 1962, No. 1, pp. 69–80. A. M. Khusro and A. N. Agarwal, *The Problem of Co-operative Farming in India*, London, 1961.

tional framework within which agriculture operates. Without these, the technical reforms offered by industry in the form of new equipment, better seed and chemical fertiliser, may not be successfully adopted. Agreement has also been widespread concerning the desirability to implement some kind of land reform which would attempt to create viable holdings by consolidating fragments and eliminating the exploitative nature of landlordism. Beyond this, however, opinions diverge sharply, some favouring more socialist forms of co-operative or collective farming, others rejecting these in favour of smaller scale, private agriculture as the best solution. Detailed empirical studies of actual examples of socialised agriculture (as opposed to somewhat sterile theorising about the nature of peasant psychology) have so far been limited to the Soviet Union and Eastern Europe.[1] These have provided the main source of evidence upon which the case for socialisation could be debated. Even to those who are not only politically uncommitted to small scale, privately owned peasant farming but, on the contrary, are impressed by the theoretical economic case for some kind of collective organisation in agriculture, the Soviet experience, at least, must be discouraging.

But to Asian economies, weighing the merits of alternative agricultural development programmes for institutional reform, the Soviet case alone is insufficient as a guide to policy. The doubt may legitimately remain that it has enough special features to permit much of the failure to be discounted when considering its application to Asia. Viewing the European cases as history it is possible to point to the ill-conceived moves or the apparently abnormal sequence of bad weather which can be held responsible for destroying sound plans. These examples of socialisation, it could be claimed, do not prove that a modified version of collectivisation would not produce the desired growth in Asian agriculture. An evaluation of the results of agricultural socialisation, even after many years is, in

---

[1] For a detailed analysis of Soviet agriculture before 1949, see N. Jasny, *The Socialised Agriculture of the U.S.S.R.*, Stanford, 1949. Much information is also contained in N. Jasny, *Soviet Industrialisation 1928–52*, Chicago, 1961. For a post-Khrushchev examination, L. Volin, *Soviet Agriculture Under Khrushchev*, American Economic Review, May 1959, No. 2, pp. 15–32.

any case, very difficult to make since so many variables enter the performance of agriculture. Widely different overall conclusions are to be drawn when evaluating the success of agricultural policies by apportioning different weights to the weather, lack of essential resources, overcentralised planning, socialist ownership, the level of taxation or farm product prices. The facts of history can be marshalled to support many programmes for development.

Much more relevant to Asian countries than European experience, and yet almost entirely undocumented, is the Chinese attempt since 1954 to promote agricultural development through socialisation. Western scholars have hardly begun to analyse either the institutions set up or the statistics of agriculture's achievements. Many difficulties attend such studies: the virtual suspension of access to China, the lack of reliable statistics and problems of definition well-known to scholars of the Soviet economy; the great expenditure of effort needed to obtain few results from the Chinese sources. Reports of brilliant successes attained by China during the First Five Year Plan period (1953–57), culminating in the " Great Leap Forward " of 1958, and followed by natural disasters, planning failures, and acute food shortages between 1960 and 1962 have, therefore, not yet undergone scholarly analysis. Certain generalisations, however, have been widely accepted as correct, but usually they were originally made as a comparison with the Soviet economy. This comparison is both necessary and interesting but it can obscure the real impact of events in China itself. For example, it has been generally argued that the socialisation campaign in Chinese agriculture had a much smoother passage than its Soviet counterpart. Fewer head of livestock were killed by the peasants as a protest while the towns did not suffer from food shortages through the peasants' refusal to deliver the planned volume of crops. In expressing such views, for example, Professor A. Nove[1] recently argued that one reason for less resistance to collectivisation in China than in Russia, was that the pig—the principal domestic

---

[1] A. Nove, *Collectivisation of Agriculture in Russia and China*, in E. F. Szczepanik, *University of Hong Kong Proceedings of the Symposium on Economic and Social Problems of the Far East*, London, 1962.

animal—was left in private hands " at this stage " (which pre-
sumably refers to 1956). Professor Nove, concluding that the
mass slaughter of livestock was avoided, stated: " . . . there
appears to have been no appreciable effect on production."[1]
Professor A. Eckstein[2] has made similar generalisations to these,
again focussing on the comparison with the Soviet Union. In
relation to the dislocation of Soviet agriculture the focus on
China's success is undoubtedly valid. But here the danger of
emphasising the comparison is illustrated. Later, evidence will
be presented to show that this view of the Chinese livestock
position during collectivisation is an oversimplification which
conceals a serious crisis, with widespread repercussions, both
on the organisation of agriculture and on the economy as a
whole.

Plan implementation and the difficulties encountered by
government during this process are topics about which too
little is known. Case studies are needed which will provide
some guide concerning the possible speed with which the programme might be introduced; the nature and magnitude of
resistance; the problems which are likely to arise during the
transition period when both a socialised and private agricultural economy exist side by side; the managerial problems of
dealing with large rural planning units. This essay attempts to
explore one very limited aspect of the implementation of
socialisation in Chinese agriculture during the period 1956–
1962. It deals with the Government's policy towards the private
sector, in particular the small private plots of land used by the
peasants in the co-operatives and collectives to grow vegetables
and rear domestic animals. The Russian experience in this
sphere merited very serious consideration by the Chinese
Government before it made any attempts to eliminate these—
the " vestiges of private ownership." Both for political and
economic reasons the Soviet Government was forced to restore
the private plots which it repeatedly tried to abolish.[3] A small

[1] A. Nove, ibid., p. 19.

[2] A. Eckstein, *The Strategy of Economic Development in Communist China*,
American Economic Review, May 1961, No. 2, pp. 508–517.

[3] See N. Jasny, *The Socialised Agriculture of the U.S.S.R.* and *Soviet Industrialisation 1928–52*, ibid.

but important private sector, composed of 18 million private plots each about 0·5 of a hectare in size and supplying a remarkably high proportion of the nation's dairy produce and fresh vegetables,[1] has persisted alongside the large scale, mechanised collective farms in the Soviet Union. The plots have for long been described by the Government as a temporary phenomenon, soon to be absorbed into the socialised sector. Yet they remain.

There can be no doubt that the Chinese leaders had studied the details of the collectivisation of Soviet agriculture. Mao Tse-tung and other Government spokesmen, in their speeches on agriculture frequently emphasised the importance of learning from Russia's experience, pointing out that China would have to adopt its own, unique solution. The actual policies adopted in China towards the private plots, however, the subsequent disruption and retreats by the Government, were almost identical to those of the Soviet Union. It is no exaggeration to state that Chinese policy towards the private sector of agriculture proved to be probably the most important single barometer of political and economic stability in the countryside during the period 1956–62. As an incentive for the peasants, the private plots played a crucial role.

The monograph is in three parts. Part one provides an introductory description of the institutions established during the socialisation of agriculture, from the mutual aid teams to the co-operatives, collectives and communes. The periods covered by each stage are outlined in this section, which aims primarily to give the background necessary for the main part, the discussion of the nature and role of the private sector. Part two assesses the importance of the private sector, first to the peasants, in food and income, secondly, to the Government as a source of important products and as an impediment to planning. A survey of the products of the private sector pro-

---

[1] The most detailed examination of this specific subject is: J. A. Newth, *Soviet Agriculture: the Private Sector.* Part I Soviet Studies, October 1961, No. 2, pp. 160–171; Part II, ibid., April 1962, No. 4, pp. 414–432. Newth estimated that the private sector supplied the following percentage of total produce in 1959: Potatoes 63%; Vegetables 46%; 32% of all livestock (reduced to a common unit); milk over 50%; eggs 82%.

vides an answer to the question of whether there were adequate reasons why the Government should wish to eliminate it entirely. Finally, part three traces the actual movement of policy towards the private sector, 1956 to 1962, in the context of the state of planning in the socialised sector of agriculture and to a lesser extent the performance of the economy as a whole. It must be stressed that though reference is made to some of the problems and achievements of collectivised agriculture in China this is not an economic appraisal of collectivisation. The focus is on the private sector and its contribution. The aim of the essay is to outline one of the many real problems encountered by the Chinese Government in carrying through a social and economic revolution in the countryside.

PART ONE

# The Background to Socialisation

# 1

# The Socialisation of Chinese Agriculture: Timing and Institutions

This chapter traces the chronology of socialisation in agriculture and describes briefly the institutions established, from the mutual aid team to the commune. Detailed treatment of this important subject would require an entirely separate study. This is merely an outline, seeking to provide the background necessary for the subsequent discussions of the economic significance of the private sector and of government policy towards it throughout the years 1956-62. It concentrates, therefore, on the changing position of the private sector in the regulations of the institutions.

Throughout the 1950's agriculture was the foundation of the Chinese economy. The success of the plans for industry, commerce, consumption or investment all depended largely upon its performance. Out of a total population which grew, perhaps, from 580 million in 1953 to 700 million in 1962, 85% were rural. More than 80% of consumer goods and almost all the food and clothing consumed in China originated directly or indirectly in domestic agriculture. Roughly 80% of light industry's raw materials, 70% of total exports, 50% of government revenue[1] and, in 1952, 50% of gross domestic product[2]

[1] Yang Ch'i-hsien, *On the Theme the National Economic Plan must Allocate according to the Sequence Agriculture, Light Industry, Heavy Industry.* Dagong bao (Impartial Daily, Peking. The Chinese newspapers referred to are published in Peking unless stated otherwise). December 11th, 1961, p. 3. To avoid the lengthy printing of Chinese titles in romanised form they have been translated into English. All references to Chinese works, therefore, are to works in the Chinese language unless otherwise stated.

[2] A. Eckstein's calculation for 1952 in his, *The National Income of Communist China*, New York 1962, p. 74.

stemmed from the same source. The exact burden to be carried by agriculture was not indicated until the First Five Year Plan,[1] (1953–57) was published in July, 1955. It set very high targets for industrial expansion. For example, it planned to expand the gross value of output of producer goods by 128% (more than double its 1952 level); to produce 4 million tons of steel in 1957 (involving a rise of 206% on 1952); to raise cement production by 110%. To meet the demand for food and industrial crops implied by these targets the plan aimed to increase the gross value of agricultural production by 23% in 5 years, equal to a compound rate of 4·3% a year. Apart from institutional impediments, the agricultural situation was characterised by a ratio of 0·287 hectares of arable land per head of total population in 1952 (it had fallen to 0·264 hectares by 1957);[2] primitive farm implements (in 1958 only 2% of the arable area was mechanically ploughed);[3] virtually no domestic output of chemical fertilisers, while the water conservation problem was probably unrivalled in the world.

It was a fundamental tenet of the Chinese Government that the technical revolution required in agriculture was only possible if preceded by the socialist reform of its institutional framework. A planned, socialised industry required a planned, socialised agriculture. Until Autumn 1955, however, every statement made by the Party or Government on institutional reform in agriculture stressed the importance of introducing socialism only over a long period. Such reform must wait until the ideological level of the peasants had been raised. Recognising the " deep attachment " of the peasants for their land, these statements affirmed that private land ownership would dominate rural China for many years.

During the few months of 1955 following the bumper harvest which claimed 175 million tons of food grains alone, these long term goals for socialist reform in agriculture were

[1] *The First Five Year Plan for the Development of the National Economy* 1953–57. Peking, 1955.

[2] Wang Kuang-wei, *Views on Allocating the Agricultural Labour Force.* Jihua jingji (Planned Economy, Journal of the State Planning Committee) 1957, No. 8, p. 6.

[3] Editorial, *Speed up the Development of the Livestock Industry*, Renmin ribao (People's Daily) February 23rd, 1959, p. 1 and 3.

abandoned. In swift succession semi-socialist agricultural producers' co-operatives and socialist collectives were established throughout China. It was then that the limits within which a private sector (in the socialist framework) would be tolerated became a crucial issue, both for the Government and peasants. Consequently, it is the regulations regarding the economic organisation of the co-operative, collective and commune—the socialist institutions—and in particular those regulations pertaining to the private sector, which are the prime concern of this chapter. To put these in their historical context, however, the reforms of the preceeding years, 1950–54, must first be outlined.

1. 1950–54: *Land reform and Mutual Aid Teams.*

The land reform[1] redistributed 46 million hectares of land among 300 million poor and landless peasants. Landlords as a class were eliminated; the rich peasants " polarised ". The campaign concentrated on winning the support of the poor and middle peasants, especially the latter who were well endowed with implements, draught animals and managerial skill. The success of the land reform in achieving its aim to raise agricultural output to pre-war levels largely depended on them. In 1952 the Government claimed that for many crops this level had been achieved. It now began to prepare the way for more basic institutional reform in agriculture, by formalizing and extending the traditional practice by which several households pooled their labour during the busy seasons. One of the most serious problems impeding agricultural development was how to raise labour productivity by reducing the seasonal fluctuations in demand for labour. Periods of almost complete

---

[1] The basic documents for the land reform are (1) *Land Reform Law of the Chinese People's Republic.* June 28th, 1950, reprinted in *Collection of Selected Laws of the Chinese People's Republic.* Peking, 1957, pp. 127–136. (ii) *Resolution concerning the Division of Rural Classes,* August 4th, 1950, reprinted in *Collection of Selected Laws* etc., ibid., pp. 136–160. Good secondary sources are Liao Lu-yen *The Great Success of the Three Year Land Reform Campaign,* September 21st, 1952, reprinted in *New China's Economic Achievements of the past Three Years,* Peking, 1954, pp. 111–118. See also Hsu Ti-hsin, *An Analysis of the National Economy of China during the Transition Period,* Peking, 1957, Chapter VII, pp. 135–164.

unemployment (80% of which according to J. L. Buck's survey,[1] was concentrated into the 4 months between November and February) were followed by periods when labour shortage reduced production. In its simplest form the mutual aid team,[2] which consisted of 7–8 households, pooled labour only during peak seasons. At the end of the season any debts outstanding between members were settled. In every other respect agriculture was managed on an individual, private basis, each household controlling the disposal of its own produce. To begin to solve the problem of seasonal unemployment and labour shortage the Government attempted[3] to make mutual aid teams operate on a permanent, all year basis. They were urged to co-ordinate agricultural and subsidiary activities, developing the latter (especially livestock rearing) in the slack months of winter. In addition they were to build up a small, collectively owned stock of implements and draught animals, financed by a levy on each member equal to 1–5% of the value of his output each year. The long term goal was to transform such teams into " land co-operatives ", in which the land, though remaining under private ownership, would be pooled and managed as a single unit. By the end of 1954[4] over 60% of all peasant households in China were grouped into mutual aid teams. Only 45%, however, were members of permanent teams, while 11% were in " mutual aid co-operatives ".

2.  1955: *Semi-socialist Agricultural Producers' Co-operatives.*

During the first half of 1955 the theoretical case for co-operativisation was stated with increasing frequency. In fact

[1] J. L. Buck, op. cit.

[2] Mutual aid teams are discussed in *An Analysis of the National Economy of China* etc. op. cit. A great deal of information is contained in Chinese Academy of Sciences Economics Research Institute, *Compendium of Materials on Agricultural Producers' Co-operatives during the Reconstruction Period of the National Economy* 1949–52, Vol. 1, Peking, 1957.

[3] Central Committee *Resolution on Agricultural Producers' Mutual Aid Co-operation*, February 15th, 1953, reprinted in *Compendium of Materials* etc., op. cit. Vol. 1, pp. 3–14.

[4] State Statistical Bureau *Statistical Materials on Agricultural Co-operativisation and the Distribution of the Product in Co-operatives during* 1955. Peking, 1957. Reprinted in Xinhua banyuekan (New China Semi-monthly), Vol. 94, 1956, No. 20, pp. 63–65.

two cases were made: a positive and a negative one. Apart from ideological reasons for desiring the socialisation of the means of production, the positive case rested on three main points.

(*a*) Larger planning units would operate with lower costs.

(*b*) The reduction in the number of producing units and strengthening of planning would enable the Government to increase its control over the distribution of the crops; in other words it could more rigidly control the level of consumption (and investment).

(*c*) The pooling of land would eradicate perhaps the main structural weakness of agriculture: the fragmentation of small, uneconomic holdings.[1] It is probable that neither the land reform nor the mutual aid teams more than touched this immense problem. Fragmentation not only wasted land, but impeded the adoption of better management and agreement on projects to extend irrigation and water conservation.

The negative argument was that unless socialism could be firmly established in the countryside, capitalism would take over. Already, it was claimed, the old inequalities of ownership, income and exploitation which existed before the land reform were reappearing. Poor peasants, lacking capital and managerial experience, had failed to maintain their independence on becoming landowners. They had fallen into debt and sold land to the rich peasants who were their creditors. In contrast the latter were combining farming with peddling.

In February 1955, Liao Lu-yen, Minister of Agriculture, revealed[2] that during the first two years of planning (1953 and 1954) the targets for cotton, silk, tea, jute, oil crops and live-

---

[1] In Buck's survey of 16,700 farms in 22 Provinces during 1929–33 the average number of fragments per farm was 5·6; 20% of farms had 6–10 fragments. The situation was exacerbated by the fact that, on average, each farm had 11·6 fields. Buck, op. cit., p. 181.

[2] Liao Lu-yen, *Report on the Basic Condition of Agricultural Production in 1954 and Measures to Raise Present Production*, March 3rd, 1955. Reprinted in *Rural Work Questions of* 1955, Peking, *1955*, pp. 10–23.

stock had not been fulfilled.[1] The planned purchase and supply
scheme[2] for many crops, so vital to successful planning, had run
into difficulties. In some areas buying agencies had failed to
deliver their required volume owing to evasion by the growers,
while in other places too much grain had been mobilised,
leaving insufficient in the hands of the growers for seed, fodder
and domestic food consumption. Liao only partly blamed the
natural disasters for the failures of 1954, emphasizing the point
that bad planning had been an important factor. The growing
" contradiction " between agricultural and industrial growth
was said to have revealed the inadequacy of the existing
institutions of agriculture to promote the required progress.
And yet the dangers of rapid socialisation were clearly indicated
by the strains already encountered in the 600,000 co-operatives
newly established. Liao Lu-yen listed them. Premature
attempts to socialise livestock at low prices had resulted in
slaughter; there was confusion and argument about the
amount of rent to be paid to landowners; cadres had been
guilty of the extremes of " commandism " (rigid implementation of plans regardless of local views) and laisser faire.
Despite the urgent need for a change in institutions, official
policy at this stage, as expressed by Liao, favoured a cautious,
long term policy of reform.

The First Five Year Plan was the first document to be
published which outlined[3] the precise time-span envisaged for
the socialist reform of agriculture. The plan aimed to
strengthen the mutual aid teams in order to prepare for co-
operativisation during the Second Five Year Plan period
(1958–62). By the end of 1957, 33% of peasant households
in China were to be in semi-socialist co-operatives (50% in the

[1] Figures published later revealed the following decline in output 1953-54:
Rice 0·5%; Miscellaneous Grains 3%; Potatoes 2%; Soya Beans 9·5%.
[2] This had begun in November 1953. *Directive on Implementing the
Planned Purchase and Planned Supply of Grain,* November 19th, 1953. Reprinted in *Collection of Selected Laws* etc., op. cit., pp. 441–443. For details
of difficulties so far encountered and proposals to improve the working of
the scheme see Central Committee *Directive on Speeding up the Re-organisation
of the Work of the Unified Marketing of Grain,* April 28th, 1955 published in
*Rural Work Questions of 1955,* op. cit., pp. 24–26.
[3] *First Five Year Plan,* op. cit., Chapter IV, pp. 79–93.

North and East, in view of their longer history of communist influence). The transition to socialism was to take 15 years. Introducing the Plan to the National People's Congress in July 1955, Li Fu-ch'un chairman of the State Planning Committee, condemned examples of " vulgar and speedy " co-operativisation in 1953 and 1954. He insisted that in future the " voluntary principle " must be rigidly applied when establishing co-operatives. The targets for agricultural socialisation in the Plan, however, were cast aside as soon as they were announced. On July 31st, Mao Tse-tung addressed[1] Party secretaries from all over China on the question of co-operativisation. The policy of consolidation and cautious transformation, advocated by Liao Lu-yen and Li Fu-ch'un, was dismissed in favour of a new socialist mass movement in the countryside. Mao likened some of his comrades to women with bound feet, vacillating between one position and another, complaining that the pace was too quick. In spring 1955 the Central Committee of the Party had set a target to increase the number of co-operatives from the current 650,000 to 1 million by October 1956. Mao now amended the target to 1·3 million. By October 1956, every *hsiang* in China (excluding border areas) was to have examples of co-operatives. But despite the call for a new campaign, Mao's plan for complete co-operativisation, in relation to what actually happened, appears moderate. By the end of 1957 or Spring 1958, 250 million people, or 55 million peasant households, were to be in co-operatives. Mao stated that by then semi-socialist reform of agriculture would be complete in some *hsien* and *sheng*, while " a small proportion " of co-operatives would have been changed into collectives. By 1960, the co-operatives would embrace the remaining 50% of peasant families and throughout the Second Five Year Plan period collectives would increase in number. The " high tide of socialist construction," which was about to begin, was " inevitable " according to Mao. Figures published later showed the speed with which it spread. At the end of March 1955 the co-operatives had embraced 14% of households, by the end of October 32%, and at the end of the year

[1] Mao Tse-tung, *On the Question of Co-operativisation.* Xinhua yuebao. (New China Monthly), Vol. 73, 1955, No. 11, pp. 1–8.

63%.[1] The Party's resolution on co-operatives,[2] October 1955, and more specifically the " model regulations for agricultural producers' co-operatives "[3] published in November set out in great detail how the co-operatives were to be organized. The regulations were a remarkable attempt to provide a set of rules which could cover almost every possible problem. The clauses revelant to this discussion must now be described.

The main characteristic of a co-operative was that member households agreed to pool their land, the use of which would be planned centrally by the co-operative's management committee on a basis of targets handed down by the Government. The co-operative paid rent to the owners, its level depending on the quantity and quality of the land. The regulations stipulated that rent should neither be so high relative to the reward for labour as to reduce the incentive to work in the fields, nor so low that owners of productive land were unwilling to place it in the hands of the co-operative. The co-operative allocated to each household a small " retained plot " for private use. It was given freely and without an obligation to pay rent. The published aim of this land distribution was to " pay attention to the peasants' need to grow vegetables and other garden crops," but in fact its main purpose was to help to make the change in land management more acceptable to the peasants. The size of the private plot was to be fixed according to the number of people in the household and the amount of arable land in the village. As the plots were to be allocated out of the genuine arable area this policy represented a real reduction in arable managed by the co-operative. However, precisely because the plots were in the fields it was a simple matter to absorb them into the co-operatively managed economy. At its maximum the area given per head must not exceed 5% of the average arable area per head in the village.[4]

---

[1] *Statistical Materials on Agricultural Co-operativisation,* op. cit.

[2] *Resolution on the Question of Co-operativisation,* October 11th, 1955. Xinhua yuebao, Vol. 73, 1955, No. 11, pp. 9–13.

[3] *Model Regulations for Agricultural Producers' Co-operatives,* November 9th, 1955. Xinhua yuebao, Vol. 74, 1955, No. 12, pp. 141–149.

[4] The Party's Resolution of October 1955 had proposed that the area of plot per head should be between 2% and 5% of the arable land per head. The model regulations dropped the 2% minimum.

Not only land remained privately owned but also farm implements, draught animals and trees. If they were hired or entirely managed by the co-operative, a payment for their use was to be made to the owners. The immediate aim was to bring draught animals, large implements and groups of trees into public ownership wherever agreement was possible between the owner and co-operative. While preserving the principle of mutual benefit, co-operatives were urged to make such acquisitions at fair prices. It was not the aim to acquire domestic animals, small tools and scattered trees.

The co-operative controlled the distribution of the produce, the regulations prescribing the method which should be followed. At the end of each harvest the first task was to ensure that obligations to the Government—taxes and sales of crops to government buying agencies—had been met. " Costs of production " were then deducted, along with the co-operative's investment funds (about 8% of the total value of production). Only then were payments made to owners of land and other resources used by the co-operatives. The residual was the sum to be distributed among members for their labour. The aim was to decentralise this task in the co-operatives in such a way as to leave the sub-divisions (the teams) enough autonomy to maintain local incentives at a high level. In order to ensure that income was closely related to productivity, a contract system between the co-operatives and teams was to be implemented. A peasant's labour was rewarded on a basis of the number of " labour days "[1] he earned. Grain ration levels

---

[1] The " labour day " system was a method of allocating points to each member for work done. The labour day was a double measure: (i) it was an index of quality and quantity of work done by a member in the collective economy; (ii) it was the unit for allocating to each member the share of production due to him, as a reward for his labour. Relative shares thus depended on the number of labour days earned by each member. Wherever possible every job was given a " norm " or target; for example a norm was set for the area of land to be ploughed by a member in a working day. Each norm was fixed by considering the performance of an " average " worker in a working day. After setting the norm the reward, in labour days registered on fulfilment, was decided with reference to the skill and strength required, as well as the importance of the job in production. An " average " norm earned one labour day, or usually 10 work points (gongfen). The

(continued p.12)

were set for different sections of society by the Government, and these rations were distributed throughout the year. " Income " earned above the value of the grain ration was to be paid in cash. The final reckoning of the income due to a peasant was carried out at the end of the Autumn harvest.

The co-operative, therefore, was only semi-socialist in character. The private sector was still an important source of income. Control over the use of almost all factors of production and the annual output, however, had passed into the hands of the co-operative. It remained to be seen how long the payments to owners of resources would continue.

3.   1956-7: *Collective;* 1958: *Communes.*

In December, 1955, Mao Tse-tung stated[1] that by the end of 1956 co-operativisation would be complete. He still gave 1959-60, however, as the date for collectivisation. These targets like their predecessors were abandoned as soon as they were set. Official figures claimed that at the end of 1955 only 4% of households were in collectives. In February 1956 the figure was already 51%, in June 63% and in December 88%.[2] Model regulations for collectives[3] were issued during June 1956. Collectives were bigger and more socialist than the co-operatives. Land now became the property of the collective, without compensation. The payment of rent to the previous owners ceased. Draught animals, large implements and groups of trees were also to be collectivised at agreed prices " according to local values ". The private plot, allocated as in the co-operative on the 5% criterion, still remained in the collective.

(Continuation of footnote 1—p. 11.)

value of a labour day earned by a youth or woman worker was less than 10 points since their norms were less than the " average ", which was based on the performance of an adult, skilled male. Members' points were registered after each day's work. See *Model Regulations*, op. cit. Also Chiang Hsueh-mo, *The Distribution System of Socialism*, Shanghai, 1962, especially Chapter 6, pp. 81–98.

[1] This was in his introduction to: General Office of the Central Committee of the Chinese Communist Party *The Socialist High Tide in the Chinese Countryside*. 3 volumes, Peking, 1956, Vol. 1, pp. 1–4.

[2] *Statistical Materials on Agricultural Co-operativisation etc.*, op. cit.

[3] *Model Regulations for Advanced Stage Agricultural Prdoucers' Co-operatives*, June 30th, 1956. Xinhua banyuekan, Vol. 88, 1956, No. 14, pp. 19-25. To distinguish these from the true co-operatives they will be called collectives.

Similarly, domestic livestock (mainly pigs and poultry), scattered trees and small implements were left in private hands. The method of distributing the product remained essentially the same as in the co-operative, but an attempt was made to improve the piece-work system for labour working in the fields.

Two years later, during the Summer of 1958, the collectives were merged into communes, which differed from the collectives in four major respects.[1]

(i) As a planning unit, the commune was much bigger both in size and scope. It was concerned with the co-ordination of every type of activity: agriculture, industry, education and defence. Its predecessor, the collective, was merely an agriculture unit.

(ii) The *hsiang* government was merged with the commune administration.

(iii) Food consumption no longer depended entirely on the amount of work done. A percentage was given freely to each person in the commune, regardless of whether he or she worked. Commune messhalls were set up to facilitate the distribution and consumption of food. The free supply proportion, therefore, represented the application of one of the first principles of communism: distribution according to need. The rest of a person's food consumption was still related to the work he did.

(iv) The private plot, land beneath houses and all trees were communised. Small numbers of domestic animals might still be privately reared but with the abolition of the private plot it is difficult to see how this was possible.

The socialist transformation of agriculture was thus carried out during three short periods of rapid change. Table I illustrates these periods.

[1] *Draft Regulations of the Weihsing People's Commune*, Renmin ribao, September 4th, 1958. Reprinted in Xinhua banyuekan, Vol. 139, 1958, No. 17, pp. 59–62.

## TABLE I

## The Three Bursts of Socialist Transformation in Chinese Agriculture, 1955–58

Percentage of all peasant households involved

| Periods of Rapid Change | Mutual Aid Teams | Agricultural Producers' Co-operatives | Collectives | Communes |
|---|---|---|---|---|
| End of 1954 | 60 | 10.9 | | |
| End of June 1955 | 65 | 14.2 } +45% | | |
| End of Dec. 1955 | | 59.3 } | 4.0 ⌉ | |
| II  End of Jan. 1956 | | 49.6 | 10.7 ⌐ +47% | |
| February, 1956 | | | 51.0 ⌡ | |
| June, 1956 | | 28.7  63.2 | | |
| December, 1956 | | 8.5  87.8 | | |
| Spring, 1958 | | 99.7[a] | | |
| III  End of Aug. 1958 | | | | 30.4[b] } +68% |
| End of Sept. 1958 | | | | 98.2[a] } |
| End of Dec. 1958 | | | | 99.1[b] |

SOURCES:

[a] From figures in *All the Nation's Villages have Virtually Carried out Communisation*, Tongji gongzuo (Statistical Work, Journal of the State Statistical Bureau) 1958, No. 20, p. 23.

[b] *Ten Great Years*, State Statistical Bureau, Peking, 1959, p. 36.
All other figures are from *Statistical Materials on Agricultural Co-operativisation* etc., op. cit.

Table II summarises the salient features of the institutions established during the years 1954–58. As it is an attempt to generalise, it inevitably hides many interesting and important variations existing in management and size. An exploration of these, however, is outside the scope of this study.

NOTES AND SOURCES TO TABLE II

[1] *Compendium of Materials on Agricultural Producers' Co-operatives during the Reconstruction period of the National Economy, 1949–52*, op. cit.

[2] These are all China averages, from figures in *Statistical Materials on Agricultural Co-operativisation* etc., op. cit.

[3] See *Model Regulations for Agricultural Producers' Co-operatives*, op. cit. Details of regional variations are found in *The Socialist High Tide in the Chinese Countryside*, op. cit.

[4] Average for 26,733 co-operatives drawn from all over China. The results of the survey are published in *Statistical Materials on Agricultural Co-operativisation* etc., op. cit.

[5] *Model Regulations for Advanced Stage Agricultural Producers' Co-operatives*, op. cit.

[6] *Directive on Strengthening the Production Leadership and Organisation and Construction of Collectives*, September 12th, 1956. Xinhua banyuekan, Vol. 93, 1956, No. 19, pp. 53–59. These recommendations were repeated a year later in two directives of September 14th, 1957.

[7] Average for the 24,249 collectives in Hopei Province, see Li T'ie, *Lessons from the experience of carrying out Collectivisation in Hopei Province*. Xinhua banyuekan, Vol. 94, 1956, No. 20, pp. 64–66. Li T'ie was First Secretary of the Provincial Party Committee.

[8] *Materials of a Model Survey concerning the Distribution of the Product in 228 Collectives during 1957*. Xinhua banyuekan, Vol. 140, 1958, No. 18, pp. 94–97.

[9] Liao Lu-yen, *Tasks of the Agricultural Front in 1959*. Hongqi (Red Flag) 1959, No. 1, pp. 11–18. Notice the marked decline in size since 1957.

[10] *Draft Regulations of the Weihsing People's Commune*, op. cit. Also *Central Committee Resolution on questions concerning the establishment of the People's Communes*, August 29th, 1958. Renmin ribao, September 10th, 1958.

[11] *All the Villages of China have already basically carried out Communisation*, op. cit. Kwangtung and Peking had an average of 9,000 and 11,000 households per commune respectively; Kweichow had 1,400.

[12] Liao Lu-yen, *Strive vigorously, attain an Abundant Harvest*. Hongqi 1961, No. 3–4, pp. 26–32.

[13] For example see a report from Hunan in Renmin ribao, May 28th, 1959; figures for Kansu in Renmin ribao, November 10th, 1960; a Renmin ribao editorial of November 25th, 1960 gives some figures for Shensi.

## TABLE II

## The Structure of Agricultural Planning Units, 1952–1962

| Period | Planning Unit Predominating | Scope and Management | Size of Planning Unit Number of Households | Ownership of Resources | Distribution System |
|---|---|---|---|---|---|
| 1952–Mid 1955 | Mutual Aid Team | Agricultural work of individual farms carried out by member households pooling their labour. Some all year, but most seasonal teams. A few with unified management.[3] | 1952:[3] 6<br>1955:[3] 8 | Entirely private in most teams. Some teams with a small, collectively owned stock of implements and draught animals. | Taxes and sales due to the State met by individual households. Debts between team members settled by assessing labour done. |
| Mid 1955–Spring 1956 | Primary Stage Agricultural Producers' Co-operative | Agricultural work, subsidiaries and capital formation centrally managed by co-op., according to plan from Government. Co-op. run by Committees and subdivided into production teams, equivalent in size to mutual aid team.[3] | Mid 1955: 28<br>Autumn 1955: 30<br>End 1955: 40 | Land private—rent paid by co-op. for its use—but managed centrally by co-op. Same for most draught animals, and implements. Co-op. aim to buy livestock and implements at agreed prices. Private plot allocated to members at 5% rule. 1956—first attempt to squeeze private plot. | Product distributed by co-op. according to principles laid down by Government. Deductions first for taxes and deliveries to State, followed by costs, reserves, rents (land and livestock etc). Residual distributed among members according to labour days earned.[4] Contracts made with teams to preserve local differentials and incentives. |
| Spring 1956–Aug. 1958 | Advanced Stage Agricultural Producers' Co-operative (Collective) | Similar to co-operative. Three tiers of management: collective, production brigade, production team.[6] | *Collective:*<br>Rec'mended by G'ment:[6] 100–300<br>Average for Hopei Prov.:[7] 340<br>Average for 228 Collectives:[8] 337<br>Average for China in 1958:[9] 170<br>*Brigade:*<br>Rec'mended by G'ment:[6] 20–40<br>Average for 228 collectives:[8] 27<br>*Team:* . | All major factors of production collectively owned. Land collectivised without compensation; other resources at agreed prices. Private plot retained as in co-op. Mid 1957; allocation criterion raised to 10%. Spring 1958, second drive against private plot. | As in co-op., except that rent no longer paid for use of land or other resources. |

## TABLE II continued

| Period | Planning Unit Predominating | Scope and Management | Size of Planning Unit Number of Households | Ownership of Resources | Distribution System |
|---|---|---|---|---|---|
| Aug. 1958–Spring 1959 | People's Commun | Co-ordinated planning of all activities in commune area, from agriculture, industry, commerce to defence and education. Fusion of *hsiang* Government and commune administration. Layers of management in commune: commune, large production brigade, production team.[10] | *Commune*:[11] End Sept. 1958: 4616 Modal Group: 5000–8000 *Large Brigade*:[12] 250 *Brigade*:[12] 30–40 *Team*:[13] 6–20 | Communisation of all means of production, including private plot, and all except a small number of domestic animals. | As in collective, with following important difference: Food consumed per household no longer entirely dependent on work done. A percentage now given *free*, per *head*, distributed and eaten in the commune messhall. Remainder of food available for consumption distributed according to labour days, after a short-lived attempt to introduce a wages system.[14] |
| Spring 1959–End 1960 | Large Production Brigade of Commune | Large brigade given more powers; made basic unit of management. Commune level cadres moved to large brigade. Tasks contracted out to different layers of management.[9] | Large Brigade:[9] 100–350 | Call for a return of private plot, and more private livestock, especially pigs. More powers of ownership given to large brigade, rather than commune.[14] | Movement away from egalitarian distribution both between and within villages. Reduction in percentage of food distributed as free supply. Large brigade given more control over distribution.[13] |
| After 1960 | Production Brigade of Commune | Further decentralisation. More cadres moved down to basic level production units. Greater powers for brigade and team in agricultural planning.[16] | Brigade:[17] 20–30 | Return of private plot.[18] Main source of pig supply to be private. Powers of ownership given to brigade and team. | As in large brigade. Lower levels of management given more powers over distribution. Disappearance of free supply. |

17

[14] Detailed discussions are available in Niao Chia-p'ei, Ch'en Liu-yuen and others. *Attempted Discussion on the Revolution of the Rural Distribution System in the Communisation Campaign.* Jingji yanjiu (Economic Research, Journal of the Chinese Academy of Sciences) 1958, No. 10, pp. 1–7. This suggests that as much as 50% of food consumed should be supplied freely. Luo K'eng-mo, however, in *On the Supply System*, Jingji yanjiu 1958, No. 11, pp. 1–8, argues in favour of 30–40% as the appropriate proportion.

[15] This was called for by the Central Committee of the Party in *Resolution Concerning certain Questions affecting the Communes,* December 10th, 1958. Xinhua banyuekan, Vol. 146, 1958, No. 24, pp. 3–11. A spate of articles on how the different provinces were implementing the directives is to be found in the national press during Spring, 1959.

[16] For example see Renmin ribao, November 10th, 1960; November 25th, 1960; December 21st, 1960; December 30th, 1960; June 21st, 1961; Gongren ribao (Workers' Daily) July 26th, 1961 (the third article in a very informative series on the communes); Remin ribao, June 16th, 1962.

[17] Gongren ribao, July 26th, 1961.

[18] As this will be considered fully in part three, references will be given in the text.


Most of the contents of Table II have already been discussed. One feature requires comment. An examination of the compartments in Table II which describe management and size reveals that the process of institutional change involved superimposing larger planning units on to those already existing. The co-operative of 30–40 households replaced the mutual aid team of 5–8 households, but the co-operative itself was divided, for actual work, into production teams equal in size to the former mutual aid team, although the powers and obligations of the production teams, of course, were different from those of the mutual aid team. Similarly the collective, with its 100–300 households, was organized on a three tier system: collective, production brigade and production team. The bridage was now the same size as the entire co-operative which had been abolished. In the commune, the large brigade[1] was the same size as the former collective. By building upon the existing structure in this way, the Government maintained considerable flexibility in rural planning. When excessive centralisation in the communes during 1958 gave rise to local dissatisfaction and demands for greater autonomy at the village level, it was easy to retain the formal structure of the commune while making the actual unit of profit and loss, ownership, day to day planning and distribution the former collective or even co-operative. In this respect it was (intentionally or otherwise) an ingenious institutional structure. Its flexibility was put to the test during the decentralisation in the communes after 1959.

[1] *Shengchan da dui* has been translated as " large production brigade ", the term *shengchan dui* as " production brigade " and not as " production team ", the usual translation. The Chinese *zu* has been translated as " team ". In the Chinese sources there is so much confusion in the use of these terms that it was considered necessary to keep as close to the Chinese as possible to maintain clarity.

PART TWO

# The Economic Significance of the Private Sector of Agriculture

# 2

# The Economic Significance of the Private Sector of Agriculture to the Peasants

Much could be written about the psychological importance of private ownership to peasants, but this chapter considers its practical importance—the food and cash which it might and possibly did provide. The private sector was bound to be important to the peasants because of the security it offered, partly in the form of estimable food supplies but to a greater extent in the form of cash from sales of produce. The amount of food forthcoming from the co-operative or collective each year could not be accurately estimated by the peasants in advance. Grain rations depended not only on prescribed levels handed down by the Government but also on local production. If production fell or the Government decided to raise the rate of saving, rations could easily be reduced. The fulfilment of work norms and subsequent credit of labour days (or work points) indicated little about the *value* of the labour day or about the final allocation of income in cash or kind. These depended on a number of uncertainties including the volume of agricultural output, the prices paid for crops by the State, the level of taxation and the rate of investment set in the co-operative or collective. Furthermore, it was the Government's aim that cash payments for such work as water conservation should be postponed as long as possible—a valid anti-inflationary device but adding to the peasants' uncertainty about their income level. The food supplied to peasants was more stable and estimable than cash. Whereas most cash payments due might be withheld until after the Autumn harvest (minor advances to cover essential needs had to be made), food was allocated throughout the year. Apart from essential food grains and minor cash advances (based on forecasts of final output and labour days already worked), there was considerable uncertainty, then, about the final level of income. The private sector could not eradicate this uncertainty but it did provide some security. A peasant manager could decide whether to consume vegetables from the private plot or sell them on the local free market (called " rural fairs "). He did not, however, have quite the same powers over the disposal of pigmeat or poultry. Before he was free to choose between their consumption or sale he had to sell a fixed amount to the State marketing authorities. There was always the uncertainty that the quotas and prices might be changed. Nonetheless the private sector did provide a fair measure of security to the peasant in that his efforts would, in the absence of natural disasters, be reflected in a level of production which could be roughly estimated and the use of which was partly under his control. When agriculture was collectivised the importance of the private sector must have increased considerably, since payments of rent for land, draught animals and tools hired by the co-operative ceased to be made. A study of 26,733 co-operatives in Autumn 1955 revealed that land rent alone averaged 75 yuan per peasant household during that year, or 18% of net income per household (cash and kind).

The measurement of the possible contribution of the private sector to the standard of living of the peasants has been divided into two parts: the first dealing with food supplies, the second with income.

## I. THE PRIVATE PLOT AS A SOURCE OF FOOD

Two steps were involved in estimating the volume of food obtainable from the private plot:

(*a*) Collection of available evidence regarding the actual size of plots in different parts of China.

(*b*) Estimation of the volume of food which such plots might produce. Table III summarises data on the size of plots.

## The Size of the Private Plot 1955-61

| | Date | Size of Private Plot in square metres — per head | Size of Private Plot in square metres — per household | Average Number of Persons in Household | Criterion for allocation theoretically in force | Actual criterion operating |
|---|---|---|---|---|---|---|
| Modal size for evidence drawn from many areas of China[a] | 1956 to mid 1957 | 26·8 | 84·5 | 3·2 | up to 5% | 2·5–3·0% |
| Average[b] sizes found in different broad regions: | | | | | | |
| North, N.W., N.E. China | ,, | 107·3 | 483 | 4·5 | up to 5% | 2·5–3·0% |
| West & Central China | ,, | 32·8 | 154·2 | 4·7 | up to 5% | 2·5–3·0% |
| South, S.W., S.E. China | ,, | 22·1 | 87·1 | 3·9 | up to 5% | 2·5–3·0% |
| National average size if 5% criterion applied throughout China | | 100·5 | 451·8 | Assumed 4·5 | 5% | |
| Average for 228 collectives scattered throughout[d] China | Autumn 1957 | 119 | 547 | 4·6 | up to 10% | 5·2% |
| National average size if 10% criterion applied[c] | 1957 | 195 | 878 | 4·5 | 10% | |
| Average size for 18,302 households in Liaoning[e] | Spring 1958 | 41 | 221 | 5·4 | 10% | 3% |
| Average size for 6750 households in 34 collectives of Hopei[f] | Spring 1958 | 37 | 131 | 3·5 | 10% | 2·2% |
| Chinese Estimate for all Szechuan[g] | 1958 | 59 | 295 | Assumed 5 | 10% | 5% |
| Odd observations in Anhwei[h] | 1958 | 87 | 348 Given | 4 Assumed | up to 5% | |
| 1 large production brigade in Chekiang[i] | March 1961 | 77 | 385 | 5 | up to 5% | |
| 4 households in Kwangtung[j] | August 1961 / Nov. 1961 | 124 | 875 | 7 | up to 5% | |

SOURCES:

ᵃ Calculated from figures in *Discussion on the question of the Retained Plot in Agricultural Producers' Co-operatives.* Jingji yanjiu 1957, No. 4, pp. 7–17.

ᵇ From figures in source (*a*) plus data in Xinhua banyuekan, Vol. 93, pp. 63–64, and Vol. 95, pp. 196–197.

ᶜ Population figures are from Tongji gongzuo 1957, No. 11 and Jihua jingji 1957, No. 7. Ninety per cent of total population is assumed to be rural. Arable land area is from *Ten Great Years*, op. cit., p. 113.

ᵈ *Materials of a Model Survey concerning the Distribution of the Product in 228 Collectives during* 1957, op. cit.

ᵉ *Survey Research on the Question of the Transition in Collectives to a Communal Ownership System.* Xin jianshe (New Construction) 1958 No. 9, pp. 1–7.

ᶠ *Discussion on the Features and Direction of the Communisation Campaign.* Xin jianshe 1958, No. 10, pp. 9–12.

ᵍ *Statistical Methods and Regulations must start with Considerations of Practicability and Follow the Line of the Masses.* Jihua yu tongji (Planning and Statistics) 1959, No. 3. I am indebted to Prof. C. M. Li for drawing my attention to this reference. Population figure for Szechuan is in *Ten Great Years*, op. cit., p. 9.

ʰ *The Way To Develop the Pig-rearing Industry.* Renmin ribao, March 7th, 1961.

ⁱ *Collective Production is the Main, Household Subsidiary Occupations the Auxiliary.* Dagong bao, August 2nd, 1961.

ʲ *Members' Household Subsidiary Occupations are the Auxiliary of the Collective Economy.* Renmin ribao, November 5th, 1961.

The figures in Table III are simply a collection of all the usable evidence of the actual sizes of private plots in different areas and at different times. They are in no sense a " representative " group. The 1956 evidence of large plots in North rather than in South China is to be expected. In the first place because land is more productive in South China than in the North, the area of plot required to provide a given volume of food is less. Secondly, the use of a criterion for allocating the plot that related to the area of arable land per head in the neighbourhood resulted in smallest plots where population pressure on land was greatest, that is, in South China. No example was discovered where the maximum size of plot, according to the criterion theoretically in force, was being allocated. In 1956 the actual criterion used was 2·5%–3% as opposed to the theoretical maximum of 5%. After the criterion was raised to 10% in Summer 1957,[1] the only example of plot size found for Autumn 1957 was, indeed, larger than those of 1956, but even here only a 5·2% criterion, or half the theoretical maximum, was being observed. The evidence for Spring 1958, showing smaller sizes, is consistent with the policy of the time to erode and abolish the plot. Again, the comparatively large plots found in the fragmentary evidence for 1961 (especially the remarkably large size in Kwangtung) are a reflection of the agricultural crisis and the policy of restoring the private plot as an incentive measure.

The main question arising from Table III is: What volume of food could be obtained from plots of such size? The private plots supplied a wide range of products including vegetables, tobacco, oil-bearing and fodder crops. The livestock products were mainly from pigs and poultry. In estimating the productivity of the plots contained in Table III the simplifying assumption was made that they were used only for vegetables and pig rearing (the two major uses in fact). Figures of green vegetable yields in China were assembled from five kinds of evidence:[2]

(a) Vegetable yields actually obtained from private plots.

(b) Yields recorded on collective land.

(c) Chinese estimates of land required to grow enough vegetables to feed members of commune messhalls and, in addition, give a small saleable surplus.

(d) Yields published in connection with campaigns to grow more vegetables.

(e) Independent yield data for pre-Communist China collected by the University of Nanking.

[1] *Decision concerning the Increase in Members' Retained Plot in Agricultural Producers' Co-operatives*, June 25th, 1957. Xinhua banyuekan, Vol. 112, 1957, No. 14, p. 153.

[2] For example see figures in Jingji yanjiu 1957, No. 4, pp. 7–17; Renmin ribao, December 12th, 1959, September 20th, 1960, March 7th, 1961; Dagong bao, October 30th, 1961, January 14th, 1962, May 19th, 1962. Nanking figures are in T. H. Shen, *Agricultural Resources of China*, New York, 1951.

The range in yields found in these sources was very great but there was enough consistency to allow for the selection of reasonable assumptions concerning yields of green vegetables in different regions. The assumptions adopted can be considered on the low side compared with yields obtainable in China from medium quality land, with good management. J. L. Buck[1] found that many villages grew virtually no vegetables at all, as the general level of knowledge among farmers about the techniques of growing vegetables was extremely low. Communist Chinese sources frequently make the point that better management in future can easily result in much higher vegetable yields.

For pigs, figures[2] were assembled for the amount of fodder and fodder land required per pig, together with the amount of pork obtained per pig at a given age. Evidence for these from many sources proved to be very similar, so that assumptions for the various broad regions of China could be made.

Different results for food output could be obtained by varying the assumptions regarding the proportions of plot devoted to pigs and vegetables. The simple model of Table IV assumes that if the plot was large enough (if the land required to grow fodder for one pig was less than the total plot area) at least one pig was reared. This is so for all examples of plot in Table III except one (the average plot in 6,750 households of Hopei). Relatively small plots were assumed to support one pig and larger plots two. The area devoted to pigs, therefore was first set, providing a residual area for vegetables. As already stated, not all pigmeat produced might be retained by the peasant for consumption. Virtually no evidence of delivery quotas, however, was discovered. Furthermore the number of pigs which could be privately reared was not entirely dependent on the supplies of fodder from the private plot; fodder could be purchased on the local market or from the State. Therefore the simple assumption was made that deliveries of pork were

---

[1] J. L. Buck, op. cit., especially Chapter XIV.
[2] See Jingji yanjiu 1957, No. 4, op. cit.; Xinhua banyuekan, Vol. 157, 1959, No. 11, pp. 116–117; Renmin ribao, May 20th, 1959; May 28th, 1959; December 17th, 1959; February 2nd, 1961; March 7th, 1961; March 12th, 1961; Dagong bao, July 24th, 1961, Hongqi 1960, No. 2, pp. 17–27.

fulfilled by rearing pigs through buying in fodder, so that all the pork produced from the private plot's fodder is assumed to be available for consumption. The food output per head thus obtained is given in Table IV.

The most striking feature of Table IV is the minute quantities of food which might be produced by the modal plot in the 1956 evidence. Such a plot provided 72 grams (2·5 ounces) of pork per head and 56 grams (1·97 ounces) of vegetables per head each day. It is impossible to estimate how widespread such small plots were during 1956. Assuming that they were fairly common, it is easy to explain the decline in incentives associated with government policy towards the private plot and its subsequent decision to raise the 5% limit to 10%. A standard by which the data in Tables III and IV can be judged is the minimum useful size of plot. This may be defined as a plot which would provide (a) 200 kilograms of vegetables per head per annum for consumption; (b) 250 kilograms of vegetables per household per annum for sales; (c) enough fodder to keep one pig per household. On these assumptions the minimum useful plot per head would be 110 square metres in North China and 60 square metres in South China. Most plot sizes in Table III are too small to meet these requirements, especially those existing in 1956. However, the plots in North China, 1956–7 (with 107 square metres per head) and in the 228 collectives (1957), after the 10% criterion had been introduced (with 119 square metres per head) are above the minimum useful size, supplying considerably more than the 548 grams of vegetables per day required for consumption in the above definition. The large plots existing in 1961, especially in the Kwangtung example (with 124 square metres per head and 1464 grams of vegetables per head per day) are now seen in better perspective. The Chinese Academy of Sciences,[1] calculating the size of plot needed during 1957 to provide self-sufficiency in vegetables, fodder, oil crops and a saleable surplus adequate for the household to buy oil and salt, concluded that 100 square metres per head were required in South

[1] Jingji yanjiu 1957, No. 4, op. cit.

## TABLE IV

### Estimates of Food Output from the Private Plot

| | Size of Private Plot per Household (from Table III) | Estimated Area of Land required per Pig | Assumed number of Pigs kept per Household | Estimated Pork Produced per annum | Private Plot left for Vegetables | Assumed Yield of Vegetables | Estimated Food output per head per day from Private Plot | | Calories per head per day from Private Plot |
|---|---|---|---|---|---|---|---|---|---|
| | | | | | | | Pork | Vegetables | |
| | sq. metres | sq. metres | | kilograms | sq. metres | kg. per sq. metre | grams | grams | Calories |
| Modal size for evidence drawn from many areas of China | 84·5 | 67 | 1 | 84 | 17·5 | 3·73 | 72 | 56 | 284 |
| Average sizes found in different broad regions:— | | | | | | | | | |
| N., N.W., N.E. China | 483 | 134 | 1 | 84 | 349 | 2·98 | 51 | 634 | 347 |
| W. & Central China | 154·2 | 26·8 | 1 | 84 | 127·4 | 3·73 | 49 | 277 | 285 |
| S., S.W., S.E. China | 87·1 | 20·1 | 1 | 84 | 67 | 4·48 | 59 | 211 | 272 |
| National average size if 5% criterion applied | 451·8 | 67 | 2 | 168 | 317·8 | 3·73 | 102 | 723 | 559 |
| Average size in 228 collectives scattered throughout China | 547 | 67 | 2 | 168 | 413 | 3·73 | 100 | 917 | 599 |
| National average size if 10% criterion applied | 878 | 67 | 2 | 168 | 744 | 3·73 | 102 | 1690 | 799 |
| 18,302 households in Liaoning | 221 | 134 | 1 | 84 | 87 | 2·98 | 42 | 131 | 188 |
| 6,750 households in Hopei | 131 | 134 | 0 | 0 | 131 | 2·98 | — | 305 | 75 |
| Estimate for Szechuan | 295 | 26·8 | 1 | 84 | 268·2 | 3·73 | 46 | 549 | 307 |
| Observations in Anhwei | 348 | 26·8 | 1 | 84 | 321·2 | 3·73 | 57 | 822 | 416 |
| Large production brigade in Chekiang | 385 | 26·8 | 1 | 84 | 358·2 | 3·73 | 46 | 733 | 353 |
| 4 households in Kwangtung | 875 | 20·1 | 1 | 168 | 834·8 | 4·48 | 66 | 1464 | 609 |

China and " over 100 square metres " in North China. Allowing for the difference in assumptions, the Academy of Sciences' estimate tallies with the above minimum useful size. Table III records that, if the 5% criterion for allocating the private plot is applied to national statistics of population and arable area, the average size of plot per head works out at 100·5 square metres, again similar to the minimum useful size. The choice of the 5% criterion thus appears to have been made on the basis of very reasonable assumptions. If applied it would have provided the peasants with 548 grams of vegetables per person each day for consumption (the consumption level required by the minimum useful plot), 175 grams per head per day for sale and 102 grams per day of pork. A 10% criterion was, therefore, so lavish (supplying 799 calories per head per day) that, not surprisingly, little or no evidence of plots approaching the size implied by the criterion was discovered in the course of this research. The examples for 1961 in Table III suggest a high percentage criterion in operation but it was not possible to calculate the figure.

Although it might be argued that the figure for pork available for peasant consumption is too high, in two respects the estimates of food output in Table IV were based on assumptions likely to underestimate rather than overestimate the amount produced. The yield figures adopted for vegetables were on the conservative side, while the calculation did not include estimates of food from poultry—because insufficient evidence was collected to enable the necessary assumptions to be made. In many regions its contribution to household food supplies would be important. For example, in Table IV the 4 Kwangtung households, each with 7 people, kept an average of 4–5 head of poultry.

In a diet composed overwhelmingly of cereals, the green vegetables and livestock products of the private plot were more important than their weight suggests. Vegetables, especially carrots and cabbage, are rich in vitamin A and calcium, in contrast to rice and potatoes, which have no vitamin A and minute quantities of calcium. Pork and eggs contain protein, eggs also having vitamin A. Thus these foods helped to balance a diet which had serious deficiencies. It would make a great

4

difference to diet if the higher yields of vegetables recorded in certain areas of China were to become more widespread through better management. The evidence available suggests that a doubling of yields is possible in many instances.

Finally, a superficial reference must be made to the possible supply of food from the private plot in relation to supplies of grain from the collective sector in 1956. During the discussion in China at the end of 1957 on food consumption and marketing it was claimed[1] that in 1956 on average approximately 765 grams of food grains per head per day were distributed by the co-operatives, collectives and State supply network to the rural population. The composition of this total was not given. In estimating the number of calories implied by the figure, widely differing results are obtainable depending on the assumptions chosen concerning the division of total grain supplied into fine grain such as rice, and coarse grain such as peas or potatoes. For example, rice provides 3·5 calories per gram, potatoes 0·74 calories. Here the simple assumptions were made that the percentage composition of the 765 grams available per head was the same as the composition of total grain output in 1956, as recorded in " *Ten Great Years* ", but " miscellaneous grains " consisted only of green peas. On these assumptions the grain from the collective sector would provide 1792 calories per head per day. If the modal private plot of 1956 supplied 284 calories (Table IV), then out of a total of 2076 calories, perhaps 86% came from the collective sector and 14% from the private plot. This exercise does not give a picture of the relative contribution of the two sectors to a peasant's food supply, since figures for important items of food are missing (for example, eggs, fish and fruit from the private sector, edible oil from the public sector). The figures do suggest that the private plot might be very important in raising the calorie supply well above bare subsistence. The complex question of calorie requirements and the adequacy of 2076 calories per head per day cannot be considered here. The average of 2076 calories conceals the fact that considerably more would be available for the labour force, because approximately 40% of China's population, in

[1] Chu Hang, *The Basic Condition of China's Grain this Year*. Xinhua banyuekan, Vol. 98, 1956, No. 24, pp. 71–73.

1953, was either under 16 years of age or over 60 years, and the calorie requirements of these groups (especially the old) are somewhat lower.


## II.   THE PRIVATE SECTOR AS A SOURCE OF INCOME

The food output figures in part I of this chapter referred to the private plot only. But in considering income it was not possible to separate income from the plot and that from other private agricultural activities. The figures for income from the private sector in this part include income from the plot, from horticulture (for example, sales of fruit) and from private livestock rearing. They do not include income from household handicrafts. Furthermore these income figures cannot be related to the data presented in the preceding part. The sources from which they were compiled are almost entirely different.

Meaningful and relevant income data for peasant households were found to be extremely rare in several years of Chinese sources. Income figures were indeed published but usually at least one component necessary for economic analysis was missing, or the definitions of income were impossible to decipher. The successful assembly of useful income figures leads to the equally difficult exercise of estimating the purchasing power of that income. Only a few data on peasant budgets and the cost of living have been published in China but these proved to be very interesting. The figures in this section, then, must be regarded as even more tentative than those for the physical productivity of the private plot.

Surveys were found which provided figures from which the proportion of income from the collective and private sectors could be calculated. The percentage from the private sector ranged from as little as 9% in 1956 to over 60% in 1957, with numerous examples of percentages in between. Since all the figures were not suitable for presentation in any systematic form, for illustrative purposes only, the best set has been presented. They are shown in Table V.

## TABLE V

### Peasant Income in 4231 Households During 1956 and 1957

| | | Net income per head Yuan | Income per head from the collective sector Yuan | Income per head from the private sector Yuan | % of Income from the private sector |
|---|---|---|---|---|---|
| N.W. China & Inner Mongolia | 1956 | 70·6 | 60·9 | 9·7 | 13·7 |
| | 1957 | 71·3 | 58·0 | 13·3 | 18·6 |
| N.E. China | 1956 | 83·7 | 67·2 | 14·5 | 17·4 |
| | 1957 | 77·3 | 57·8 | 19·5 | 25·2 |
| Central Plain area | 1956 | 61·4 | 50·1 | 11·3 | 18·4 |
| | 1957 | 63·7 | 50·9 | 12·8 | 20·1 |
| South China | 1956 | 65·0 | 47·4 | 17·6 | 27·1 |
| | 1957 | 68·9 | 45·7 | 23·2 | 33·6 |

Source:

From figures in *Materials of a Model Survey concerning the Distribution of the Product in 228 Collectives during* 1957, op. cit.

In Table V " net income " per head refers to disposable income received from the collective (including the value of any payments in kind), plus income from the private sector (which includes the value of produce consumed and receipts from sales). The existence of both the incentive and adequate opportunity, relative to the collective economy, to conceal some of the private output and income means that the private share should be regarded as an understatement of the actual percentage.

The level of income from the collective in Table V fell in most areas in 1957, compared with 1956. The author of the survey attributes this to the 1957 policy of raising the rate of investment in collectives. In contrast, the level of private income rose in 1956–57. This was due first to the fact that in 1956 it was very low, associated with the dislocations of the collectivisation drive; secondly, because in 1957 measures were taken to increase it, including the granting of bigger private plots and higher pig prices. Thus in 1957 the private sector increased its contribution to total income.

But to indicate the significance of these figures of income from the private sector to a peasant household it is necessary to consider the purchasing power of the yuan, or the peasants' cost of living. Table VI was compiled from data in budget studies published in China during 1957. The figures were mainly drawn from the Chinese surveys conducted in Shensi, Hunan and Fukien provinces. To combine these data into one table is to ignore regional variations both in prices and requirements of certain goods (for example, coal). However, despite the lack of refinement the table does help to make income figures in yuan more meaningful.

TABLE VI

The Cost of Living of Chinese Peasants in 1956

| Item | Assumed Consumption per head per year | Price per unit Yuan | Cost Yuan |
|---|---|---|---|
| Rice | 192·5 kg. | 0·150 | 28·90 |
| Potatoes | 66·5 kg. | 0·026 | 1·73 |
| Sugar | 0·3 kg. | 0·960 | 0·29 |
| Pork | 3·33 kg. | 0·920 | 3·07 |
| Oil | 2·00 kg. | 1·120 | 2·24 |
| Salt | 5·00 kg. | 0·300 | 1·50 |
| Cloth | 12 metres | 0·400 | 4·80 |
| Socks | 1 pair | 0·500 | 0·50 |
| Shoes | 1 pair | 4·500 | 4·50 |
| Cotton wadding | 0·25 kg. | 2·400 | 0·60 |
| Medicines | — | — | 2·00 |
| Lamp Oil | 0·5 kg. | 1·00 | 0·50 |
| Coal | 150 kg. | 0·002 | 3·00 |

Total cost per head per annum: 53·63

From figures in T'an Chen-lin: *Preliminary Research on the Income and Living Standard of China's Peasants.* Xinhua banyuekan, Vol. 109, 1957, No. 11, pp. 105–111.

The consumption quantities, costs and prices in Table VI are from actual peasant budgets. Apart from the minute pork consumption, food intake levels are moderate (food was also available from the private plot). Expenditure on clothing,[1] as conceded by T'an Chen-lin is low while the budget omits items such as tobacco, matches, haircuts, repairs and replacement of tools, all of which, according to a Fukien example, account for about 7 yuan per head each year. The Chinese author of the budget study claimed that a total income per head of 60 yuan and over could be regarded as a fair standard, but points out that the cost of living varied regionally. In mountainous areas he assessed the annual income per head needed for the kind of standard of living in Table VI as 40–50 yuan; in North China it was 50–60 yuan; Hunan, Hupei, Kiangsi and South West China it was 60–80 yuan. Comparing these needs with the figures showing the sources of income in Table V it appears that the income from the private sector could very easily be the decisive factor in the standard of living, given the need and determination of the Government to maintain high investment and taxation rates in the collectives. When basic food requirements cost approximately 30 yuan per head, in a year of low output (both on the collective and private plot) the standard of living could be precarious. After paying for food rations little would be left for clothing, fuel and medicines, not to mention less basic consumer goods. When the private plot was squeezed in 1956 the situation did, indeed, become tense in many peasant households.

Some evidence was found on the relationship between peasant class and income from the private sector. At the time of the land reform all peasants were registered[2] as belonging to

---

[1] The low cloth consumption is confirmed by the fact that in August 1957, the ration of cloth per head was reduced from 20·65 metres to 16 metres. The budget of Table VI assumes 12 metres: see Editorial, *Everyone Must Economise on Cloth*, Renmin ribao, August 20th, 1957.

[2] *Resolution Concerning the Division of Rural Classes*, op. cit., 1950, laid down the criteria for classification. A peasant's class was to depend on whether he owned land, worked land or rented it out; employed labour or worked himself; loaned money for interest, and on his general standard of living. Thus a landlord owned land, did little or no work and received most of his

four broad classes: the landlord class, rich, middle (upper and lower) and poor. After the land reform it was claimed that many former poor peasants rose into the middle class. When agriculture was collectivised, most of the distinguishing features of class, based on private ownership of land and other agricultural capital, ceased to exist. All classes received the same size of private plot per head and presumably any bias in allocation would be towards the former poor peasants. But the ownership of valuable assets not yet collectivised was still unequal. The middle and rich peasants continued to own such livestock as pigs and milk cows, as well as small farm implements and scattered trees. Did this, together with the fact that they were often better farm managers than the poor peasants, mean that the rich and middle peasants received a larger percentage of their income from the private sector than the poor peasants, and were they, therefore, more independent of the collective for their livelihood? If so this could be a matter of great political and economic significance. When the Chinese Government attacked private ownership in agriculture during 1956 and 1958 it used this as one of its most important arguments in favour of complete socialisation. Two sets of evidence concerning class and source of income are worth recording. The first is from a 1956 survey of 860 households in two collectives of Yunnan Province. The figures in Table VII were not published by the Chinese to make this particular point concerning class and source of income and in this respect they are more reliable than the second example in Tables VIII and IX.

income from rent and interest payments. A rich peasant owned and rented in land, employed labour and received some income from rent and interest. The main distinguishing feature of a middle peasant was that, although he might own land and a good stock of working capital in the form of draught animals and implements (like the rich peasants and landlords), he relied heavily on his own labour. To belong to the " upper middle ", rather than the rich peasant class, income from " exploitation " must not generally exceed 5% of the total. Where the household's living standard was depressed by such factors as large family size but small labour supply or a shortage of arable land, this proportion might be as high as 30%. Finally a poor peasant owned little or no land, sold his labour and paid large sums of interest and rent.

## TABLE VII

### Income in 860 Yunnan Peasant Households, 1956

| Class of Household | Net Income per head | Income per head from the collective sector | Income per head from the private sector | % of Income from the private sector |
|---|---|---|---|---|
| | Yuan | Yuan | Yuan | |
| Poor | 81·3 | 73·6 | 7·7 | 9·5 |
| Middle | 91·6 | 78·9 | 12·7 | 13·9 |
| Former rich | 101·9 | 84·9 | 17·0 | 16·7 |
| Former landlords | 97·8 | 86·8 | 11·0 | 11·3 |
| Former small rentiers | 90·7 | 75·9 | 14·8 | 16·4 |

Source:

Wang Min, *Several Questions seen from Income Distribution in Two Collectives*. Xinhua banyuekan, Vol. 98, 1956, No. 24, pp. 59–60.

Table VII shows that the poor peasants were heavily dependent on the collective for income but the other classes could still not be described as independent. A more comprehensive survey for 1957 on this very theme was published by the Amoy University Economics and Politics research group in 1958. This study was published precisely in order to illustrate the point that the higher peasant classes in Fukien, by owning most of those income earning assets still allowed to be privately owned, were so independent of the collective for income that the collective economy's existence was threatened by their non-co-operation. Table VIII reproduces the figures for the ownership of certain assets in two of the collectives surveyed.

## TABLE VIII

### Class Distribution of Certain Privately Owned Assets in Two Collectives of Fukien, 1957

| | Percentages of Total | | | | |
|---|---|---|---|---|---|
| | Collective A | | | Collective B | |
| Class of Peasant Household | Milk Cows | Mandarin orange trees | Peach trees | "Mulihua" (a spice) Average quality | Top quality |
| Poor peasant: | 5 | 11 | 56 | 30 | 40 |
| Lower middle: | 20 | 15 | 38 | 70 | 60 |
| Upper middle: | 75 | 74 | 6 | | |

Source:

Amoy University Economics and Politics Research Group, *The Complete Elimination of the Remnants of Private Ownership of Producer Goods is a Necessity for the Development of Productive Power*. Xuexu luntan (Forum of Learning) 1958, No. 3, pp. 21–30.

With the exception of peach trees, the food producing and income-earning assets in Table VIII were highly concentrated into the hands of the middle peasants. The authors explain the exception by the fact that capital requirements of peach growing and income from peach sales were low relative to mandarin oranges. The poor peasants, therefore, specialised in peach growing, the middle peasants preferring to grow the more costly and yet more lucrative orange.

The importance of the private sector as a source of income for different peasant classes in four collectives of Fukien is brought out by Table IX.

TABLE IX

Percentage of Net Income from the Private Sector of Different Peasant Classes in Four Collectives of Fukien, 1957

| Name of Collective | Class of Household | Percentages of income | | | Percentage of income from private sector |
|---|---|---|---|---|---|
| | | Milk Cows | Fruit | Other | |
| Ch'eng Men | Poor peasant | 11·6 | 11·4 | 18·6 | 41·6 |
| | Lower middle | — | 19·1 | 29·7 | 48·8 |
| | Upper middle | 9·6 | 27·8 | 29·9 | 67·3 |
| Ao Feng | Poor peasant | — | 3·5 | 16·1 | 19·6 |
| | Lower middle | — | 33·5 | 6·0 | 39·5 |
| | Upper middle | 5·9 | 28·7 | 7·6 | 42·2 |
| Hsieh Fang | Poor peasant | — | 34·6 | 17·0 | 51·6 |
| | Lower middle | — | 22·8 | 15·1 | 37·9 |
| | Upper middle | — | 38·1 | 2·2 | 40·3 |
| Lu Lei | Poor peasant | — | — | 43·6 | 43·6 |
| | Lower middle | — | 18·7 | 41·0 | 59·7 |
| | Upper middle | — | 5·8 | 54·8 | 60·6 |

Source: Calculated from figures in Xuexu Luntan, op. cit.

The Chinese text implies that the income from the collectives involved in Table IX is net income as defined earlier. The definition of income from the private economy, however, is not the one used in this study (i.e. the value of produce consumed plus income from sales) but only income from sales. This must be so at least for income from milk cows, for despite the fact that poor and lower middle peasants in Ao Feng collective are not credited with income from this source, elsewhere in the text, figures relating to ownership reveal that the poor peasants in that collective owned 5% of all milk cows, while lower middle peasants owned 20%. The income from milk cows, therefore, must mean cash income from sales. If this definition applies in Table IX to income from fruit and " other " (which must refer to income from the private plot proper), then the extent to which the private sector provided independence is actually understated for all classes, while the relative position of the peasant classes is somewhat distorted.[1]

Nevertheless the table establishes its point, that in three out of the four collectives the degree of independence provided by the private sector was much greater for the lower and upper middle peasant households than for the poor. An important reason for this, brought out in Table IX was the concentration of ownership of fruit trees and cows in the upper peasant classes. In four cases out of the twelve, private income actually exceeded collective income, while in five other cases it was more than 40% of the total income earned.

To summarise the conclusions of this chapter, in 1956 the size of private plot was generally well below the 5% maximum and in some areas it was very small indeed. Compared with the co-operative or collective economy it was generally a minor

---

[1] If this understatement of the importance of the private sector did occur it was counterbalanced by the faulty arithmetic in the article which gave some of the upper class households a greater share of income from the private sector than was the case. In the article, figures for all 4 collectives show the upper classes to be overwhelmingly more independent of the collective for income than the poor. Not all the percentage figures given were wrong, but mainly those for the upper peasant classes, so strengthening the point made by the article. The mistake was to take private income as a percentage of *Collective* income, rather than *total* income.

source of food and income, yet it was important to the peasants for four main reasons:

(i) It provided the kind of foods needed for a balanced diet.

(ii) In many cases it raised the standard of living from the margin of subsistence to tolerable levels.

(iii) It was a source of cash.

(iv) It gave some degree of security in contrast to the uncertainties attached to the possible level of income forthcoming from the collective.

In 1957 evidence suggests that the private plots and income from the entire private agricultural sector were larger than in 1956. As a result some peasants enjoyed considerable independence from the collective, especially the middle peasants.

# 3

## The Economic Significance of the Private Sector of Agriculture to the Government

In 1955, when the socialist drive began, the Chinese Government's aim for agriculture was to control and plan it according to the political and economic ends set out in the First Five Year Plan. Full socialisation was the ultimate goal, but for the present a private sector was allowed to exist. How serious a contradiction, to use the Marxist term, did this concession create? Was it important for the Government to eliminate the private sector as soon as possible, because it was likely to weaken the collectives and hence government control over agriculture? The answer to these questions depends on the size of the private sector, the nature of its products and their importance both in the peasants' living standard and to the collective economy. It also depends on the degree of control the Government could exercise over the private economy, especially control over the competition between the two sectors for labour and other resources. The fact that the private economy was allowed to exist does not mean that the Government considered it to be a minor problem, for the evidence in favour of abolishing it would have to be weighed against the possible disincentive effects of such action.

In this section it is argued that the Government had important practical reasons for wishing to eliminate the private sector of agriculture. Its significance to the peasants has already been examined. The evidence showed that, although the collective economy provided the bulk of their cash and food (especially calories), even moderately sized private plots commensurate with the 5% criterion could make the peasants independent enough to impede the work of the collective economy. Competition between the two sectors for labour, managerial skills and animal fertiliser did, in fact, affect collective production. Despite attempts to regulate the hours of work devoted to the plot, many cases of absenteeism from the collective are recorded in the Chinese sources. More will be said about this in part three. This chapter examines the feature of the private plot which preoccupied the Government: the fact that it supplied most of the fertilisers used by Chinese agriculture, in the form of pig manure. At July 1st, 1956,[1] approximately 70 million out of the 84 million pigs in China were in private hands. Under the regulations for collectives private pigs were not to be socialised while the private sector was to continue to supply the greatest proportion. Failure by the Government to mobilise enough of the fertiliser from this source on to the collectively owned land would have a disastrous effect on crop yields. Although it is not possible to calculate the exact composition of total fertiliser supplies, the Government repeatedly stated that pigs provided by far the greatest proportion. As long as this continued control over the private agricultural sector was a vital problem for the Government. But to abolish the private plots, in the absence of alternative supplies of fertiliser (either chemical, or animal fertiliser from collectively managed herds), would have serious consequences. This chapter attempts to give some indication of the extent to which the Government depended on the private plots for fertiliser. The discussion first digresses from the private plot and considers what supplies of chemical fertilisers were available for use by collective agriculture in relation to national requirements. Secondly, these supplies are compared with the volume of fertiliser derived from pigs. The comparison involves converting the animal manure into chemical fertiliser equivalent.

[1] *The Livestock Situation throughout China in* 1956, Xinhua banyuekan, Vol. 99, 1957, No. 1, pp. 88–90.

## CHEMICAL FERTILISERS IN CHINA

An examination of the successive targets adopted for chemical fertiliser production in China up to 1967 (Table X) demonstrates the point that, even if the targets are reached, the heavy reliance on animal fertilisers would continue for many years in the interim. The Table also shows that policy towards chemical fertiliser production changed during 1956-57. In 1956 the 548,000 tons planned by the First Five Year Plan for 1957 gave way to a target of 3·0 million tons for 1962—more than a fivefold increase. Only a year later it was further increased to 5·0–7·0 million tons, while the target for 1967 was fixed at 15·0 million tons.

TABLE X

The Upward Shift in the Target for Domestic Production of Chemical Fertilisers

| Date when Target fixed | Target Year | Planned Output in Target Year—Million Tons |
|---|---|---|
| July 1955[a] | 1957 | 0·548 |
| October 1957[b] | 1957 | 0·748 |
| September 1956[c] | 1962 | 3·0–3·2 |
| November 1957[d] | 1962 | 5·0–7·0 |
| November 1957[d] | 1967 | 15·0 |
| February 1958[e] | 1958 | 1·156 |
| April 1959[f] | 1959 | 1·3–1·5 |
| April 1960[g] | 1962 | 5·0–7·0 |
| April 1960[g] | 1967 | 15·0 |

Sources:

[a] *First Five Year Plan*, p. 36.

[b] *Jihua jingji*, 1957, No. 10, p. 7.

[c] *Second Five Year Plan*. Xinhua banyuekan, Vol. 94, 1956, No. 20, pp. 164–170.

[d] *Plan for Agricultural Development*, 1956–67, Revised Edition, Xinhua banyuekan, Vol. 120, 1957, No. 22, pp. 127–132.

ᶜ *Report on the Draft Plan for* 1958. Xinhua banyuekan, Vol. 127, 1958, No. 5, pp. 12–23.

ᶠ *Report on the Draft Plan for* 1959. Xinhua banyuekan, Vol. 155, 1959, No. 9, pp. 15–20.

ᵍ *Plan for Agricultural Development,* 1956–67, Peking. 1960.

Unfortunately, the announcement of these higher targets was not accompanied by a discussion of the plans to supply the necessary resources and of the shift in priorities that they would entail. The target of 15 million tons for 1967 must be regarded as that of a politician rather than an economic planner, reflecting national needs as opposed to possibilities. A contributor to the journal of the Planning Committee, Chi Ch'ung-mieh,[1] considered that chemical fertiliser output could only be expanded slowly during the Second Five Year Plan period. Without giving figures he emphasised that to build a nitrogenous fertiliser plant (the type most needed by China) made great demands on funds, equipment and technical know-ledge. The non-Party economist, Ma Yin-ch'u,[2] a believer in realistic planning, confirmed Chi Ch'ung-mieh's point, stating that the construction of a synthetic ammonia plant (to make ammonium nitrate) would require 220–260 million yuan, 38,000 tons of steel and 20,000 tons of high-grade mechanical equipment. Assuming[3] that Ma's costs referred to a factory producing 200,000 tons a year, the 1967 target of 15 million tons would involve the construction of 75 firms. At 175 million yuan per plant (allowing for improved building techniques), 13,100 million yuan would still be needed—over twice the sum planned for the entire constructional programme of the Ministry of Heavy Industry in the First Five Year Plan period. Ma Yin-ch'u believed that an output of 3–4 million tons of

[1] Chi Ch'ung-mieh, *Industry ought Actively to Assist the Development of Agriculture.* Jihua jingji 1957, No. 10, pp. 7–10.

[2] Ma Yin-ch'u, *My Economic Theory, Philosophical Thoughts and Political Standpoint,* Peking, 1958, pp. 21–22.

[3] The First Five Year Plan proposed to renovate and build 8 chemical fertiliser firms, each with an output around 200,000 tons per annum. See Wang Hsing-wei and Lan Chien-ch'ao, *Actively Develop the Chemical Fertiliser Industry,* Jihua jingji 1957, No. 10, pp. 11–15.

chemical fertilisers might be achieved in the Second Plan period.

Table XI shows actual output, total supplies and supplies per arable hectare in China, the latter being compared with Japan. In the earlier years of the First Five Year Plan, 65–70% of chemical fertilisers used in China were imported.[1]

TABLE XI

Output and Supplies of Chemical Fertiliser

| | CHINA | | | JAPAN |
|---|---|---|---|---|
| Year | Output million tons | Supplies: output and imports million tons | Supplies per arable hectare kg. | Supplies of Nitrogenous Fertiliser per arable hectare, kg. |
| 1954 | 0·298[a] | 0·802[c] | 9[e] | |
| 1955 | 0·332 | 1·255 | 14 | |
| 1956 | 0·523 | 1·608 | 17 | 587[f] |
| 1957 | 0·631 | 1·944 | 21 | 626 |
| 1958 | 0·811 | 2·708 | 31 | |
| 1959 1st half | 1·155[b] | 2·745[d] | 30 | |
| 1960 | 0·887[b] | — | | |
| Pl.1962 | 7·000 | | 77 | |

Notes and Sources:

[a] Output figures for 1954–58 exclude Ammonium Nitrate. Source: *Ten Great Years*, p. 86.

[b] *People's Handbook for* 1961, Peking, 1962, p. 251. Includes all types of chemical fertiliser.

[c] 1954–58 from *Ten Great Years*, p. 151.

[d] Import figure from China Association, London.

[e] Arable areas from *Ten Great Years*, p. 113; 1959 and 1962 estimated on 1957 arable area.

[f] Arable areas from *Japan Statistical Yearbook*, 1960, supplied by R. P. Dore. Nitrogenous fertiliser consumption from *Annual Yearbook of Production, F.A.O.*, 1959. Rome, 1960.

[1] Chi Ch'ung-mieh, op. cit.

Table XI shows the small supply[1] of chemical fertiliser available per hectare of arable land in China. Even with an output of 5 million tons a year (the 1962 minimum target), and allowing for no increase in arable land over the 1957 area, the amount would be only 13 per cent of the nitrogenous fertiliser alone used by Japan in 1957. The situation is worse than the figures in Table XI suggest, for the supplies per *sown* hectare are, of course, even lower. Policy in China since 1950 has been to rely overwhelmingly on raising yields of crops per unit area rather than increasing the total arable area through land reclamation. This has entailed attempting to increase the sown area of land each year, by double and treble cropping. Since each sowing requires fertiliser, the demand for fertiliser has increased greatly. In relation to the above supplies, Table XII has been compiled to illustrate the amount of chemical fertiliser the Chinese consider to be absolutely necessary in addition to organic manures.

TABLE XII

Chinese Estimates of Crop Requirements of Chemical Fertilisers in Addition to Organic Fertilisers

kg. per hectare sown

| Source | Rice | Wheat | Kaoliang | Sugar Beet | Sugar Cane |
|---|---|---|---|---|---|
| 1 | 149 | 74 | | 149 | 447 |
| 2 | 112–186 | 112–186 | 74–149 | | 223–298 |
| 3 | 48 | | | | |

| Source | Vegetables | Cotton | Jute | Tobacco | Rapeseed |
|---|---|---|---|---|---|
| 1 | 74 | 223 | 223 | 298 | 74 |
| 2 | | 112–186 | 223–298 | | 74–149 |

Notes and Sources:

[1] Chi Ch'ung-mieh, *Industry ought Actively to Assist the Development of Agriculture*, op. cit. These are described as *minimum* requirements.

[1] The figures for supplies per arable hectare (not sown area) are so small that it must be made clear that they are figures of fertiliser in the form of ammonium sulphate, etc., and not pure nitrogen, phosphate, potash.

5

2. *Rural Work Handbook*, Peking, 1957, pp. 328–333. This is ammonium sulphate only. They are recommended rates of application rather than minimum requirements.

3. Editorial, *Fertiliser is the Key in Continuing the Leap Forward in Agricultural Production*. Renmin ribao, July 19th, 1960, pp. 1–2. This figure for Kwangsi is for nitrogenous fertiliser only. The requirement is given as 0·9 kg. per 50 kg. of rice produced. The figure of 48 has been obtained by applying it to the average paddy yield during China in 1955 (calculated from *Ten Great Years*).

The figures in Table XII are not estimated requirements for maximum economic or physical returns but minimum to safe requirements for reasonably high yields. In fact they are very modest applications compared with Japan which, in 1957, had yields of rice 66% higher than China.[1] Japanese farmers are recommended[2] to apply 407 kilograms of nitrogenous fertiliser per hectare of rice and 405 kilograms to wheat land. Table XII compared with figures in Table XI for actual chemical fertiliser supplies available in China, demonstrates China's grave deficiency.

Experiments and soil analyses in China[3] have found that Chinese soils are more responsive to nitrogen than phosphate and to phosphate than potash, a result of the long term use of organic manures. A great deal of work is now being done in China on the physical returns to fertiliser application. Table XIII presents some evidence on the ability of fertilisers to increase the yield of different crops in China. The figures, of course, assume that the necessary supplies of water are available. It should be stressed that the figures are for the average additional physical production of various crops associated with the application of each kilogram of fertiliser at a certain level of application (where the level is given, it has been indicated in

[1] Chinese figure from *Ten Great Years*, p. 107; Japanese yields from *Annual Yearbook of Production*, 1959, *F.A.O.* Rome, 1960, p. 50.

[2] H. L. Richardson, *The Use of Fertilisers in the Far East*. Proceedings of the Fertiliser Society No. 41, London, 1956.

[3] H. L. Richardson, *Transactions of the International Society of Soil Science* Vol. I, 1952, Commission II and IV, Joint Meeting on Soil Fertility, pp. 222–230.

the notes). The figures do not indicate the nature of the marginal product curve for fertiliser applications.[1]

## TABLE XIII
### Average Physical Returns to Applications of Nitrogenous Fertiliser in China

Average Additional Yield in kg. per Hectare for each kg. of Fertiliser Applied

| Source | Rice | Wheat | Maize | Foodgrains in general | Rapeseed | Cotton |
|---|---|---|---|---|---|---|
| 1 | 3·6 | | | | | 0·9 |
| 2 | 4–5 | 2–4 | 6–8 | 3·0 | | 1·0 |
| 3 | | | 2–4 | | | |
| 4 | | | | 3·5 | | 1·0 |
| 5 | | | | 2·0 | | |
| 6 | | | | 3·0 | | |
| 7 | 2·06 | 1·25 | 2·93 | | 0·83 | 0·65 |

Notes and Sources:

1. Wang Hsing-wei and Lan Chien-ch'ao, *Actively Develop the Chemical Fertiliser Industry,* op. cit.

2. Ibid. Results for N. China.

3. Chang Yao-tsung, *Discussing Fertiliser Applications on Maize.* Zhongguo nongbao (Chinese Journal of Agriculture) 1962, No. 7, pp. 6–7. Average return on applications up to 560 kg. per hectare.

4. Chi Ch'ung-mieh, *Industry Ought Actively to Assist the Development of Agriculture,* op. cit.

5. *Agriculture is the Foundation: the Chemical Industry Must Greatly Assist Agriculture,* People's Handbook for 1961, Peking, 1962, pp. 251–252. Return for up to 372 kg. applied per hectare.

6. *Rear Pigs, Accumulate Fertiliser.* Dagong bao, August 31st, 1961.

7. H. L. Richardson in *Transactions of the International Society of Soil Science,* Vol. I, July, 1952. Commission II and IV Joint Meeting on Soil Fertility, pp. 222–230. Results of experiments carried out in China by the author during the 1930's. The average returns are for an application of 300 kg. per hectare.

[1] Chang Yao-tsung, *Discussing Fertiliser Applications on Maize,* Zhongguo nongbao, 1962, No. 7, pp. 6–7, provides data from which the following table can be made, showing the marginal returns in one experiment.

(continued on p. 50).

The figures show that for levels of fertiliser application far exceeding those attained by China the return in crop yield is likely to be very high.[1] So far Chinese attention has been on the physical returns involved. Virtually no information has been published on costs and revenues. One interesting fragment[2] suggests that the economic return will be high even for levels not likely to be reached for years to come. The cost of fertiliser per kilogram to the user was given as 0·154 yuan. Assuming that 149 kilograms of fertiliser are used (the Planning Committee's minimum requirement for rice in Table XII), the total cost of fertiliser per hectare is 22.9 yuan. On the conservative assumption that an extra 2 kilograms of grain are produced per kilogram of fertiliser applied, the value of extra grain would be 59·5 yuan.[3] Or, at a cost of 0·154 yuan per kilogram for fertiliser and 0·2 yuan as the price of grain, the margin is reached when the additional yield is 0·77 kilograms

[1] For evidence concerning Asia in general see P. Allan, *Fertilisers and Food in Asia and the Far East*. Span., Vol. 4, No. 1, 1961, pp. 32–35. Ed. A. Datta, *Paths to Economic Growth*, London, 1962, where a general return of 2 kg. of grain per kg. of nitrogenous fertiliser is cited. N. W. Pirie, *Future Sources of Food Supply; Scientific Problems*. Journal of the Royal Statistical Society, Series A (General) Vol. 125, part 3, 1962, pp. 399–417 adduces the figure of 10–20 kg. of wheat, rice, or maize per kg. of pure nitrogen (i.e. per 5 kg. ammonium sulphate).

[2] Wang Hsing-wei and Lan Chien-ch'ao, op. cit. p. 12. The return in future will probably be greater, since fertiliser costs are expected to fall. Both Chi Ch'ung-mieh op. cit., p. 8 and Ma Yin-ch'u op. cit., pp. 19–21 refer to the high price of fertiliser influencing the demand by collectives.

[3] 0·2 yuan per kg. was the approximate price for grain in 26,733 co-operatives during 1955. Calculated from *Statistical Materials on Agricultural Co-operativisation* etc., op. cit.

(continuation of footnote 1—p.49).

| Ammonium sulphate used per hectare | Yield of Maize per hectare | Rise in Maize output | Average Product per kg. of fertiliser applied | Marginal Product per kg. of fertiliser applied |
|---|---|---|---|---|
| kg. | Hundred kg. | kg. | kg. | kg. |
| 186 | 42·30 | 821 | 4·4 | 4·4 |
| 372 | 43·10 | 902 | 2·4 | 0·4 |
| 558 | 44·04 | 997 | 1·8 | 0·5 |
| 744 | 44·69 | 1057 | 1·4 | 0·3 |
| 1116 | 43·83 | 976 | 0·8 | -0·1 |

of grain. The evidence of Table XIII suggests that the level of fertiliser input involved in such a yield is too remote to cause any concern at present. Other costs involved in using fertiliser, such as labour and water, would have to be high to affect the return on fertiliser seriously.

Thus we return to the global requirements of China for chemical fertilisers and the supplies per hectare involved in these estimates. These figures are contrasted in Table XIV with the annual tonnage which would be required if the moderate amount of 300 kilograms was applied per hectare.

TABLE XIV

The Relationship Between Total Supplies of Chemical Fertiliser and Supplies per Hectare of Arable Land in China

| | Total Million Tons | Supplies of Chemical Fertiliser Per Arable Hectare kg. |
|---|---|---|
| 1. Supplies of Chemical Fertilisers per Hectare available at various levels of Total Supply | | |
| 1956–67 Plan[a] and a Pre Communist estimate[b] of requirements | 15 | 164 |
| State Planning Commission's estimate of requirements[c] | 10 | 110 |
| 2. Total Supplies required assuming 300 kg. Fertiliser applied per Hectare of Arable Land in China | 27 | |

Sources:

[a] *Revised version of* 1956–67 *Plan for Agriculture.* Xinhua banyuekan, Vol. 120, op. cit.

[b] T. H. Shen, *Agricultural Resources of China,* op. cit.

[c] Chi Ch'ung-mieh, *Industry ought Actively to Assist the Development of Agriculture,* op. cit.

A safe conclusion is that China will be very short of chemical fertiliser for many years.

## THE CHINESE PIG POPULATION AND ITS SIGNIFICANCE AS A SOURCE OF FERTILISER

Plans for the pig population of China are given in Table XV. It contains three examples of downward revision in targets forced upon the Government by the failure to attain targets set. The 1957 target is an interesting example. The aim for the end of 1957, set in Autumn of 1956, was for 138 million. In May 1957 it was reduced to 120 million and in July to 110 million. The 1962 target was revised in 1957 from its 1956 figure of 250 million to 180–240 million. Thirdly, the 1960 plan was unable to maintain the target for 1959 of 280 million, and this too was reduced to 243 million.

### TABLE XV
#### Changes in the Target for Pigs

| Date When Target Fixed | Target Year | Planned Pig Numbers (million head) |
|---|---|---|
| 1955 (1st F.Y.P.)[a] | 1957 | 138·34 |
| 1956[b] | 1957 | 138·34 |
| 1957 May[c] | 1957 | 120·0 |
| 1957 July[d] | 1957 | 110·0 |
| | | |
| 1956 (2nd F.Y.P.)[b] | 1962 | 250·0 |
| 1957[e] | 1962 | 180–240 |
| | | |
| 1958[f] | 1958 | 150 |
| 1959[g] | 1959 | 280 |
| 1960[h] | 1960 | 243 |
| | | |
| 1956[i] | 1967 | 240–360 |
| 1957[e] | 1967 | 300–360 |
| 1960[j] | 1967 | 300–360 |

Sources:

[a] *First Five Year Plan*, p. 88.

[b] *Second Five Year Proposals.* Xinhua banyuekan, Vol. 94, 1956, No. 20, pp. 164–170.

c Liao Lu-yen, *Summary of 1956 Agricultural Production Work and Tasks for 1957*. Xinhua banyuekan, Vol. 106, 1957, No. 8, pp. 81–88.

d Po I-po, *Report on Results of Carrying Out the Economic Plan for 1956 and the Draft Plan for 1957*. Xinhua banyuekan, Vol. 112, 1957, No. 14, pp. 28–39.

e *Revised Version of Plan for Agricultural Development*, 1956–67, op. cit. These are approximate figures: the plan called for $1\frac{1}{2}$–2 per household by the end of 1962 and $2\frac{1}{2}$–3 by 1967.

f Po I-po, *Report on the Draft Plan for 1959*. Xinhua banyuekan, Vol. 127, 1958, No. 5, pp. 12–23.

g Li Fu-ch'un, *Report on the Draft Plan for 1959*. Xinhua banyuekan, Vol. 155, 1959, No. 5, pp. 15–20.

h Li Fu-ch'un, *Report on the Draft Plan for 1960*. Renmin ribao, March 31st, 1960, pp. 2–3.

i *Draft Outline Plan for Agricultural Development* 1956–67, op. cit. This is an approximate figure. The plan called for 2–3 pigs per household.

j *Agricultural Development Plan*, 1956–67, 1960 version, Peking, 1960.


The compilation of a series showing the number of pigs in China (Table XVI) proved to be a very difficult task. The State Statistical Bureau's compendium of statistics, *Ten Great Years*, while presenting series for almost all important products was particularly secretive on this subject. It gave figures for 1957 and 1958. The former is almost certainly too high while the latter can be safely regarded as an even greater exaggeration. In other sources some contradictions were found.

TABLE XVI

The Number of Pigs in China 1952–1960

| Date | Pig Numbers millions |
|------|----------------------|
| 1952 | 89·77[a] |
| 1953 | 96·131[b] |
| 1954 | 101·718[b] |
| 1955 | 87·92[b] |
| 1956 | 84·40[b] |
|      | 97·80[c] |
| 1957 | > 110·120[d] |
|      | but < 138 |
|      | 127·8[e] |
|      | 145·895[a] |
| 1958 | 180[f] |
|      | 160[a] |
| 1959 | 270[g] |
|      | 180[g] |
| 1960 | 240[h] |

Sources:

[a] *Ten Great Years*, p. 117.

[b] At July 1st. Xinhua banyuekan, Vol. 99, 1957, No. 1, pp. 88–90.

[c] End of 1956. *Report on Achievements in Carrying Out the 1956 Plan and Draft Plan for 1957*. Xinhua banyuekan, Vol. 112, 1957, No. 14, pp. 28–39.

[d] Liao Lu-yen, *On a Basis of Agricultural Co-operativisation, with Revolutionary Zeal Strive to Realise the Plan for Agricultural Development*. Xinhua banyuekan, Vol. 127, 1958, No. 5, pp. 127–132. Liao stated that the figure was below the First Five Year Plan target but more than the 1957 target.

[f] Li Fu-ch'un, *Report on the Draft Plan for 1959*, op. cit. This was one of the many figures to be scaled down in 1959 (in this case by 11%).

[g] T'an Chen-lin, *Struggle to Bring Forward the Fulfilment of the Plan for Agricultural Development*, April 1960. Published with 1960 version of 1956–67 *Agricultural Plan*, Peking, 1960. 270 million was given as figure " for the whole year "; 180 million for the end of 1959. 180 million is also found in Liu Juei-lung, *The Agricultural Front in 1960*, Hongqi 1960, No. 2, pp. 17–27.

ʰ In March 1960, Li Fu-ch'un announced that the figures had not reached the Second Five Year Plan target. *Report on the Draft Plan for* 1960, op. cit.

The serious decline in pig numbers between 1954 and mid 1956 was associated with the socialisation drive in agriculture and the Government's policy towards the private plot. The recovery during the second half of 1956 and throughout 1957 was largely due to the more liberal policy adopted towards the private sector. These policies are examined in the final part of the essay.

Pig manure is rich in organic matter, nitrogen, phosphorus and potash. The effects of applying it to the soil are long lasting. In measuring its significance as a fertiliser pig manure has been converted into chemical fertiliser equivalent. For purposes of simplification and because nitrogen is both the manure's most important plant food and, in China, the soil's greatest need, the manure was converted into ammonium sulphate equivalent. Two kinds of data were used to arrive at a reasonable conversion ratio:

(i) Chinese analyses of the nitrogen content of pig manure. These were used along with a conversion ratio between nitrogen and ammonium sulphate plus figures for the volume of manure produced per pig each year.[1]

(ii) Chinese figures specifically relating the volume of manure per pig to an equivalent in ammonium sulphate.[2]

[1] Chu Kwang-ch'i, *Research on the Fertilising Effects on Farm Fertiliser*, (i) *The Fertilising Effects of Farm Fertiliser on Rice.* Turang Xuebao (Chinese Journal of Soil Science) 1959, No. 3–4, pp. 180–188. Hsi Ch'eng-fan (Deputy Director of the Chinese Academy of Sciences Soils Team) *Pig Manure is a Good Fertiliser.* Renmin ribao, December 30th, 1959, p. 7. Editorial, *Pigs are First in the Six Animals.* Renmin ribao, December 17th, 1959, p. 1. Rural Work Handbook, op. cit., p. 322. *The State of World Chemical Fertiliser Production and Consumption.* Jihua jingji 1957, No. 10, pp. 26–27.

[2] For example, in Renmin ribao, February 23rd, 1959 and December 17th, 1959; Hongqi, 1960, No. 2, pp. 17–27; Dagong bao, August 31st 1961.

The two sets of data were not always entirely reconcilable. On the whole, data in (i) were more favourable to the nitrogen content of manure than data in (ii), but there was considerable overlap in the final figures obtained from the two sets. Much of the difference could be explained by the definition of manure used (in particular, whether it included straw or mud). Estimates of the amount of ammonium sulphate equivalent to the manure of one pig in a year calculated from sources under (i), ranged mainly from 37 kilograms to 75 kilograms, with a concentration around 37–48 kilograms. Estimates under (ii), however, ranged between 30 kilograms and 50 kilograms. To be on the conservative side the figure of 30 kilograms per pig was assumed. At the end of 1955 the pig population stood at 87–92 million. This population would produce manure each year equivalent to 2·586 million tons of ammonium sulphate, or 24 kilograms per hectare of arable land. The 1955 supply of chemical fertiliser per hectare of arable, we recall, was 14 kilograms. Between July 1954 and July 1956 the pig population fell by 17·318 million. This involved a loss equivalent to 510,000 tons of ammonium sulphate or 4·6 kilograms per arable hectare. Finally, the fact that 70 million out of the 84·4 million pigs in China on July 1st, 1956 were privately owned meant that 2·064 million tons of ammonium sulphate equivalent per year, or 18·7 kilograms per arable hectare, were in private hands and most of this was required for the collective land.

The Government was, therefore, faced with a difficult situation: how to treat the private sector in such a way as to secure the maximum co-operation from the peasants and guarantee the economic progress required in collective agriculture. Control over both labour and livestock manure was essential and yet either too much control, or insufficient financial incentive could result in the slaughter of livestock, refusal to deliver manure or grain, and withdrawal into a subsistence economy wherever possible. How the Government treated this problem between 1956 and 1962 forms the subject of the final part of this study.

PART III

# Socialisation and the treatment of the Private Sector of Agriculture 1956 - 1962

# 4

## 1956 - Autumn 1957 : The First Round

Within a period of six years (1956–1961), government policy towards the private sector of agriculture had gone two full circles: two attempts to eliminate it each followed by a restoration. The attacks on the sector were associated with the socialisation drives of 1956 and 1958; the withdrawals with economic and political crises which arose shortly afterwards. The chronology of events is examined by dividing the period 1956–1962 into three:

    (i) 1956—Autumn 1957: the first round.

    (ii) 1957—end of 1960: the Great Leap Forward, elimination of the private plot, and growing agricultural crisis.

    (iii) 1961–62: decentralisation in the communes and the restoration of a private agricultural sector—in theory and in practice.

The " Great Leap Forward " of 1958 has tended to obscure the fact that 1956 was also a year of intended leap, in many ways similar to 1958. It was a year of rapid acceleration in the drive for industrial and agricultural development. In January Chou En-lai announced[1] this policy along with the corresponding high targets. The gross value of industry was to rise by 18·6% on 1955; basic constructional investment by 60%. The

---

[1] Chou En-lai, *Political Report to the Chinese People's Government Consultative Committee.* Xinhua banyuekan Vol. 79, 1956, No. 5, pp. 11–20. Chou referred to 1956 as a year of " leap " and " speed up ".

target for the volume of grain output was set at 9% over the 1955 level, while cotton output was to rise by 18%.

In agriculture the tone was further set by the targets contained in the 1956–1967 Plan[1] also published during January 1956. The plan's 40 clauses called for a technical revolution in agriculture, including the extension of double-cropping, irrigation, water and soil conservation, fertiliser application, better seed strains, the elimination of pests and animal diseases. Yields of crops were to be doubled in many areas. It will be recalled that the Government's policy on co-operatives and collectives had already been set out during 1955 in a series of important speeches and policy documents. Li Fu-ch'un, Chairman of the State Planning Committee, in July 1955 had attacked[2] the speed and compulsion of the co-operativisation programme in 1954 and 1955. Mao's statement[3] of July 1955, while clearly calling for a co-operativisation campaign, had also demanded persuasion, voluntary change and caution, arguing that the only sure way to obtain a rise in agricultural production after socialisation was through voluntary co-operation by the peasants. The Communist Party's resolution on the co-operatives,[4] October 1955, repeated these views, stressing the importance of winning over the rich and middle peasants to accepting socialist agriculture, for they were not only the owners of high quality tools and livestock but also experienced farm managers who secured high production levels from the land. With a high income from subsidiary activities they had a lot to lose by joining collectives and therefore, it was argued, persuasion, not force, was needed to avoid political and economic upheaval. The October resolution condemned co-operatives for their treatment of the private sector. Some had failed to allocate private plots, while others had allocated less than the minimum requirements. Payment of excessively low

---

[1] *Draft outline of the Plan for the Development of Chinese Agriculture*, 1956–67. Renmin ribao, January 26th, 1956. Reprinted in Xinhua banyuekan Vol. 78, 1956, No. 4, pp. 2–5.

[2] Li Fu-ch'un, *Report on The First Five Year Plan for the Development of the National Economy, July 5th*, 1955, Published with *First Five Year Plan*, op. cit.

[3] Mao Tse-tung, *On the Question of Co-operativisation*, op. cit.

[4] *Resolution on the Question of Co-operativisation*, op. cit.

prices for tools and draught animals when they were collectivised, failures to pay rent and interest on resources loaned, or cash for animal manure delivered, were other practices condemned.

The 1956–1967 Plan now confirmed the " voluntary principle " and underlined the importance of the private plot as a source of vegetables and animal fodder. Its proclaimed targets to co-operativise Chinese agriculture during 1956, followed by collectivisation in 1957, stood in stark contrast to the fact that collectivisation was already well underway, with 51% of farm households in collectives by February 1956. As an inducement to join, the plan promised that in 1956, 90% of peasant household members would have their living standard raised. The superiority of socialist agriculture in providing material benefits was to be the main factor in achieving a voluntary transformation in the countryside.

Yet, the first half of 1956 witnessed the implementation of policies towards the private sector diametrically opposed to these published government and Party instructions; it is not known what secret orders were given to the cadres. The degree of encroachment upon the private sector was not revealed until Summer 1956. At the meetings of the National People's Congress (June) and Central Committee of the Communist Party (September) the economic situation in China was thoroughly examined. Ministers, Provincial Party Secretaries and others criticised their own planning methods and failures in speeches[1] which were remarkable for their frankness and pragmatism. Condemnation of the practices destined to reduce incentives below the danger level was accompanied by restatements of policies to be followed. New measures to deal with the serious situation were introduced throughout the Autumn.

Evidence presented in part two showed that in 1956 the size of the private plot was usually less than the prescribed

---

[1] The speeches were all published in Xinhua banyuekan Vols. 88 and 89, passim (the proceedings of the National People's Congress, held in June 1956). Xinhua banyuekan Vols. 94 and 95, passim., recorded the meeting of the Central Committee of the Party.

limit of 5% of the arable land per head in the village. But, in
some areas, the plot was completely abolished by over zealous
cadres. Elsewhere, poor quality land, distant from the peasants'
households, was allocated for short periods only. This was to
reduce the peasants' interest in the plot. The regulations
relating to the ownership of resources were similarly disregarded. Trees, implements, livestock (including poultry)
were forcibly collectivised, in some cases without payment and
in others at excessively low prices. Failures to pay rent or
interest due to peasants were manifold. Loans were not repaid
on time or the period of repayment was " lengthened ";
manure duly delivered to the collective economy was not paid
for—or was undervalued. In some co-operatives and collectives
the use of labour on the private plot was so restricted that it
was impossible to farm it. Peasants who worked on the plot
more than the party cadres considered appropriate were fined
a number of labour days. Investment loans were arbitrarily
levied. Some peasants were afraid to sell produce from their
private plot in case the income earned would be channelled into
collective investment shares.[1] Social isolation by the " dubbing
of capitalist hats " was a common way of deterring peasants
from working the private plot.

Peasants responded by continuing to kill pigs and poultry,
while allowing their draught animals to die through neglect, so
exacerbating the decline which had begun during 1954 in
anticipation of the co-operativisation campaign and the uncertainty it brought regarding ownership. In Hunan this slaughter
was said[2] to be " a serious phenomenon," and it was undoubtedly on a large scale throughout China. Equally important was
the failure of peasants to renew their pig herds or to plan to
increase the numbers kept. This was not entirely due to the
policy towards the private plot. Fodder was scarce; prices
received by peasants for the pigs were considered low in
relation to costs and also to prices which had to be paid for

[1] Ma Yin-ch'u, *My Economic Theory, Philosophical Thoughts and Political
Standpoint*, op. cit., pp. 46–47.
[2] Chou Hsiao-pan, *Several Problems in Strengthening the Co-operatives*,
Xinhua banyuekan, Vol. 95, 1956, No. 21, pp. 196–197. Chou was First
Secretary in Hunan.

pork, while the tax payable on killing pigs was too high. Peasants' income, especially cash income, was seriously affected by this reduction of private activity. In one Hunan collective[1] of 209 households, income from private livestock rearing fell in 1956 by 26% on 1955; the number of privately owned pigs fell by 70%. A survey carried out in an entire *hsien* of Hunan found[2] that the decline in income from the private plot was so great that eight households classified as poor peasants had insufficient cash to buy their food rations. A further five households could only afford to buy potatoes. Many instances of peasants' lack of money to buy necessary salt and oil were documented at this time. As a result, stocks of these, other consumer goods and farm tools were accumulated by local marketing agencies. Indebtedness increased among families which had hitherto been solvent. Finally, sales to towns of produce such as vegetables, poultry and pork began to decline.

Although this catalogue of coercien and peasant responses implies a rapid decline in incentives to produce and sell, the situation would have been less serious if the economic performance of socialised agriculture and of industry had been better. In fact agriculture was suffering from planning chaos and mismanagement, leading towards crisis conditions which would soon affect the entire economy unless drastic action was taken. One of the main problems was excessive centralisation in agricultural planning. Unrealistically high targets, to be fulfilled in too little time, were handed down to the local areas by the higher planning authorities. The co-op and collective cadres implemented these plans mechanically, despite protestations from experienced local peasants that they were quite unsuitable for the area. Advice was either disregarded or not sought at all: the cadres demanded plan fulfilment. This policy was called " commandism " by the Communist Party. Thus dry land was changed to wet; single crop land to double; new seeds and close planting were introduced; tools and implements were bought. All these were technical reforms called

[1] *With much Activity in Subsidiary Industry, why aren't there better Real Benefits*, Xinhua banyuekan, Vol. 93, 1956, No. 19, pp. 63–64.
[2] Kao Chang-jen, *After Developing Many Kinds of Activity*, Xinhua banyuekan, Vol. 92, 1956, No. 18, pp. 78–79.

for by the 1956–67 Plan. Many cadres, however, introduced them without adequately considering the full implications in that area—often with bad results. Statistics for the Province of Chekiang[1] showed that to change one *mou* of land (about one fifteenth of a hectare) from single to double crop rice raised the demand for labour by 80%. Similarly, it increased the demand for draught livestock and fertiliser. In many areas where these demands could not be met the attempts at technical reform were unsuccessful. New seeds introduced from outside the region, often without trials, failed. Wells were needlessly dug when geological surveys would have revealed that irrigation water from this source was unsuitable. Peasants did not know how to use the agricultural implements purchased. Indeed, this story alone merits an article in its own right, as an example of the extent to which different departments in the government administration planned in isolation from one another, illustrating the costs of over-centralisation. Estimates[2] of the demand by agriculture for equipment after collectivisation proved to be far too high. In[3] 513 collectives scattered throughout 13 *sheng*, double-wheeled ploughs, planting machines, reaping machines. donkey boilers and water carts remained largely unused, Peasants referred to " hanging ploughs " and " sleeping carts ". Lack of technical men, servicing shops, spare parts, knowledge of how to work the machines, were all presented as reasons for this waste. Some collectives had actually been forced to buy implements.[4] The Government criticised cadres for planting too much food grain, compared to raw material crops needed by light industry. Reports coming in by June 1956 indicated[5] that the output of the oil-bearing rapeseed was 14% below the

[1] Wang Kuang-wei, *Views on the Allocation of the Agricultural Labour Force*, Jihua jingji, 1957, No. 8, pp. 6–9.

[2] For an analysis of the demand estimates for double-wheeled ploughs see: *Why has the Demand for Double-wheeled Ploughs fallen, and their Production been Stopped?* Jinhua jingji, 1956, No. 9, pp. 1–4.

[3] Editorial, *Use the Large Agricultural Implements Already Owned Properly*, Renmin ribao, March 25th, 1957.

[4] *Why has the Demand for Double-Wheeled Ploughs fallen, and their Production been Stopped?* op. cit.

[5] Teng Tzu-hui, *The Situation in the Year-old Co-operativisation Campaign and Future Work*, Xinhua banyuekan, Vol. 88, 1956, No. 14, pp. 57–61.

planned target. Sown areas for ground nuts, jute, sugar and silk were less than planned. Mulberry trees had been cut down to transfer the land to grain. Cotton output for the year was later shown[1] to have fallen by 4% on 1955, reaching a level 19% below the 1956 target.

Many difficulties arose in the collectives over the system of distribution. Excessive concentration of this task into the hands of the collective level institutions, together with the adoption of an egalitarian policy, led to dissatisfaction among hitherto rich and productive villages. Some cadres made this worse by attempting to increase the size of the collective, further separating productivity and reward. Changes in the estimated value of the labour day made by the collectives in the light of more knowledge concerning what the harvest would be, acted as a disincentive to work on the collective land. One collective in Hunan reduced the value of the labour day by 46% when it became apparent that the plan adopted was unrealistic and that the collective's disposable income would be lower than expected. With rents abolished, private plots squeezed, pig prices too low for profit and non-repayment of debts, there was a growing crisis of confidence in the collectives' ability to fulfil the promise to increase the income of 90% of peasants. This decline in expectations accelerated as the bumper harvest hoped for failed to materialise. The advent of severe natural disasters in the form of typhoons, droughts and floods, affecting 7·1% of the arable area[2] of China, further drained human energy and resources already overworked.

Finally, government spokesmen revealed that collective agriculture had so far failed to offset the decline in private livestock. The rapid rise in demand for draught animals, associated with technical reforms in agriculture, was in many areas matched only by declining numbers, and an increase in weak, diseased stock—incapable of pulling the new type of ploughs. According to the State Statistical Bureau,[3] the number of oxen

---

[1] Po I-po, *Report on the Results of Carrying out the Economic Plan for* 1956 *and the Draft Plan for* 1957, Xinhua banyuekan, Vol. 112, 1957, No. 14, pp. 28–39.

[2] Po I-po ibid.

[3] Xinhua banyuekan, Vol. 99, 1957, No. 1, op. cit.

actually increased by 1·5% during the first half of 1956, while that of water buffalos fell only by 0·5%, but these figures are difficult to reconcile with numerous local reports of grave shortages of draught animals. The State Statistical Bureau did, however, indicate that in Honan Province the situation was deteriorating. The burden of arable land per beast had risen from 21·6 *mou* in 1954 to 25·7 *mou* in 1956, compared with the levels normally considered to be the maximum for draught animals (20 *mou* for cattle and 15 *mou* for mules). Men were being substituted for livestock in ploughing. According to official figures, the 13·6% decline in pigs throughout China during 1955 (Table XVI Chapter 3) was followed in the first half of 1956 by a further 4% drop.[1] The Government readily admitted that its own grain marketing policy was partly responsible. In Autumn 1955, it had mobilised too much grain from the co-ops and collectives, leaving inadequate amounts for seed and feed. Price policy and the treatment of the private sector were the other causes. Attempts by the collectives to rear pigs on a large scale were often unsuccessful. Collectives built pig sties of lavish dimensions at costs which bore no relation to the income obtainable. One example quoted[2] by

[1] This decrease, again, seems to be an understatement, considering the Government's repeated focus of attention on the pig shortage and its attempts to remedy it. For example, in Siaokan Special District of Hupei Province, there had been a dramatic reduction, as the following figures show.

| Year | Number of Pigs | Percentage Decline on Previous Year |
|------|----------------|-------------------------------------|
| 1953 | 909,946 | |
| 1954 | 745,548 | — 18% |
| 1955 | 503,844 | — 32·4% |
| End of June 1956 | 415,346 | -- 17·5% |

Source:

*An attempt to Discuss the Question of Pig Fodder.* Liangshi Gongzuo (Grain Work) 1956, No. 21, pp. 22–25. Thus by the end of June 1956, over 54% of the pigs in existence during 1953 had disappeared in this Special District.

[2] Chin Cheng, *A Wasteful Pig Farm with Great Losses*, Jihua jingji, 1958, No. 3, p. 38.

the State Planning Committee involved a pig farm for 280 head. The investment in sties alone was four and a half times the investment in an equivalent area of domestic housing. Costs of producing pork were 56% higher than those of private rearers. Management was so bad that, despite the luxurious buildings, the death and disease rate was very high. Many cases of this kind were recorded. The failure of public and private pig herds to increase affected soil fertility, food supplies and exports (pigmeat and bristles).

There is no doubt that the Government realised the urgent need to deal with the crisis of confidence in the collectives. This is clearly brought out by the voluminous statements of senior officials condemning the above failures and policies. Both high level planners and local cadres were blamed for the situation, but on the whole the latter were exonerated on the grounds of lack of experience: their demoralisation would have been catastrophic for the Party. Many documents were published restating the correct policy towards the private sector and collectives. Collective cadres were directed to implement the regulations for collectives concerning the private plot and to settle the competition between collective and private interests voluntarily. More incentives were also to be provided for the collective economy.[1] There were to be smaller collectives and production brigades, less centralisation and more powers for the brigades. Egalitarianism was to be replaced by a system linking payment more closely to production. Taxation and grain deliveries for 1956 were to remain at their 1955 level. In September, Chou En-lai announced[2] the opening of a free market, which was intended to stimulate village trade and private sideline production. Li Hsien-nien[3] called for an increase in the prices of pigs and industrial crops.

---

[1] See, for example, the important Directive of September 12th, 1956. *On Strengthening the Building of Organisation and Control over Production in the Co-operatives*, Xinhua banyuekan, Vol. 93, 1956, No. 19, pp. 53–59.

[2] Chou En-lai, *Report on the Second Five Year Plan for the Development of the National Economy*, Xinhua banyuekan, Vol. 94, 1956, No. 20, pp. 35–49.

[3] Li Hsien-nien, *Make our Prices Still Higher; Promote the Development of Production*, Xinhua banyuekan, Vol. 95, 1956, No. 21, pp. 84–87.

Despite these changes the difficulties continued throughout the Autumn. Urban demand for food greatly exceeded the plan —since urban population had itself risen by 2.3 million in 1956 as opposed to the target of 840,000.[1] Official figures claimed that total food grain output in 1956 rose by 4·4% in the face of natural disasters and rapid socialisation, but the rising demand necessitated the reduction of food stocks. Directives[2] called for reductions in urban food rations as grain sales greatly exceeded purchases by the Government buying agencies. Collectives were criticised for distributing more grain to members than the national interest would allow. In Autumn the Government conceded[3] that many collectives were not able to provide a rise in income for 90% of peasants. Figures[4] for 19 collectives in Honan show that the incomes of 38% of former upper middle peasants, 25% of poor, 28% of former lower middle and 33% of rich peasants remained unchanged. Eighteen per cent of former lower middle peasants received a cut in income.

The 1957 plan[5] called for retrenchment in construction and consolidation in industry. A great effort was to raise agricultural production while avoiding the excesses and mistakes

[1] Po I-po, op. cit.

[2] See (i) Chu Hang, *The Basic Condition of This Year's Grain*, Xinhua banyuekan, Vol. 98, 1956, No. 24, pp. 71–73; (ii) Directive, *On Present Grain Marketing and the Work of Unified Grain Purchase and Sale after the Harvest*, October 12th, 1956, Xinhua banyuekan, Vol. 96, 1956, No. 22, pp. 89–90; (iii) Directive, *On Present Grain Work*, November 22nd, 1956. Xinhua banyuekan, Vol. 98, 1956, No. 24, p. 70; (iv) Editorial, *Speedily Control the Volume of Grain Sales*. Renmin ribao, November 24th, 1956; (v) Directive, *On Certain Concrete Problems in Distributing the Autumn Harvest in Co-operatives*, November 24th, 1956. Xinhua banyuekan, Vol. 98, 1956, No. 24, pp. 56–58.

[3] Directive of November 24th, ibid.

[4] T'an Chen-lin *Preliminary Research into the Income Position and Standard of Living of Chinese Peasants*. Xinhua banyuekan, Vol. 109, 1957, No. 11, pp. 105–111.

[5] See Speeches at Chinese People's Political Consultative Conference, February, 1957, especially by Ch'en Yun and Ch'en Cheng-jen. Xinhua banyuekan, Vol. 105, 1957, No. 7, passim. *Also* Po I-po, *Report on Carrying out the Plan for 1956 and Draft Plan for 1957*, op. cit.

of 1956. Reports[1] of the dissolution of collectives, the exodus of manpower and draught animals from agriculture into peddling and trade, spurred the Government on to make further conciliatory gestures. Teng Tzu-hui,[2] a member of the Central Committee of the Chinese Communist Party, urged *hsiang* party secretaries in May 1957 to win the support of the middle peasants—even to the point of " not violating their interests." He asked collectives to pay higher entry prices for collectivised resources, to leave trees (for example in Kwangtung) in private hands, and to adopt profit sharing schemes where convenient. Attacking the wastes of collective livestock rearing enterprises, Teng called for the return of livestock, especially pigs, to private hands. The view held by comrades who considered that this was a backward step—against the collectivist principle of socialism—was described by Teng as " mistaken, dogmatist and subjectivist." The guiding principle in policy was the benefit to production. Finally, Teng announced that the Government had issued a directive raising the limit on the private plot per head in some areas to 10% of the arable land per head in the village—a doubling of the existing limit. This concession was made general throughout China in June.[3] In Spring 1957, pig prices were increased[4] by 14% on average, along with price increases for oil bearing crops. Certain retail prices affecting mainly the town population were raised to counter this redistribution of income in favour of the peasants.

[1] For example: Ma Yin-ch'u, op. cit., pp. 103–194 et. passim. The New China News Agency, March 25th, 1957 (English) reported dissolution of co-ops. in Kwangtung involving 80,000 households after the Autumn harvest of 1956. Editorial, *Strongly Develop Peasant Trade.* Renmin ribao, November 22nd, 1956. Directive, *On Making a Good Job of Spring Ploughing to Achieve an Abundant Harvest in Agriculture during* 1957. March 19th, 1957, Xinhua banyuekan, Vol. 106, 1957, No. 8, pp. 70–73.

[2] Teng Tzu-hui, *On Internal Contradictions of Co-operatives and Democratic Management.* Xinhua banyhuekan, Vol. 109, 1957, No. 11, pp. 94–100.

[3] Standing Committee of the National People's Congress, *Decision on Increasing Members' Retained Plot in Co-operatives,* June 25th, 1957. Xinhua banyuekan, Vol. 112, 1957, No. 14, p. 153.

[4] Ministry of Finance Communique, *On Lowering the Slaughter Tax Rate, Remitting Tax Levies on the Livestock Industry,* March 2nd, 1957.

*Also* Directive, *All Areas Raise the Prices of Live Pigs and Pork,* March 1st, 1957. Both are in Xinhua banyuekan, Vol. 105, 1957, No. 7, p. 87.

The Spring of 1957 was the time of the " Hundred Flowers " bloom. For a brief period unbridled criticisms of the Party's methods were encouraged. It marked the high point of liberalism in China during the 1950's. The increase in the private sector in agriculture was in line with the political atmosphere. A very conciliatory attitude was adopted. A member of the Academy of Sciences[1] argued that private plots were quite consistent with socialist ideology; contradictions were bound to exist, but increases in income and peasants' socialist enlightenment would solve them over time. For the present, the private plot could satisfy demands which the collective could not. It reduced the administrative burden on the collective in this way and allowed it to concentrate on more important matters. The plot ensured that full use was made of small areas of land and of auxiliary labour. Competition between private and public interests for labour and fertiliser could be settled amicably through a correct income and price policy. It was 1961 before these arguments were heard again.

---

[1] Sung Hai-wen, *Discussion of the Question of the Retained Plot in Co-operatives*, Jingji yanjiu 1957, No. 4, pp. 7–17.

# 5

## Autumn 1957 - Winter 1960 : The Great Leap Forward, Communisation and Growing Agricultural Crisis

Within a year, the Chinese countryside had witnessed the Great Leap Forward and another socialist revolution with the setting up of people's communes. The private plots—no longer regarded as an important auxiliary to the socialist agricultural economy, but as a serious contradiction—had been abolished.

These policy changes were greatly influenced by the continued relative stagnation of socialised agriculture throughout 1957, in spite of the emphasis on financial incentives already described. In 1957 the gross value of agricultural production increased by 3·5% but the volume of food grain output by only 1%.[1] The sown area of food grains fell by 55 million *mou*, accounting for a decrease of 6 million tons of grain.[2] Natural disasters were said to have affected a very large area of land but these, according to Peking, were insufficient to account for the poor performance of agriculture. Industrial crops again

---

[1] *Ten Great Years*, p. 105.

[2] Liao Lu-yen, *On a Basis of Co-operativised Agriculture, with Revolutionary Zeal, Strive to Achieve the Plan for Agricultural Development.* Xinhua banyuekan, Vol. 127, 1958, No. 5, pp. 127–132.

failed to meet demand.[1] Soya bean output was down by 1·5% on 1956; cotton sown area in September was reported to be 8·3% below 1956. Through an increase in yields of 21%, however, cotton output was claimed to have reached a level 11·4% above 1956. This was still not enough to prevent the cloth ration from being cut in August 1957, following the 5% decline in cotton harvested in 1956 and 12·5% decrease in cloth output during 1957. The livestock situation continued to be serious. As in 1956, grain distribution problems were encountered in the Autumn.[2] The State marketing authorities sold more grain (especially in the towns) than their lower purchases warranted. In the collectives, peasants demanded decreases in grain taxes and sales to the State. Consequently, many collectives reduced their investment rate, distributing more grain for consumption. Grain and other crops were sold illegally on the free markets. Black markets and rising food prices (especially for vegetables) were other symptoms of the economic ills of the country.[3] The collectives, in fact, were by no means securely established in the countryside. The four features which distinguished the collective from the co-operative were (i) their greater size; (ii) their collective ownership of land, no rent being paid to members; (iii) their collective ownership of draught animals; (iv) the higher investment rate. These were

[1] Liao Lu-yen, ibid.

Also T'an Chen-lin, *Explaining the Second Revised Edition of the Development Plan for Agriculture, 1956–67.* May 17th, 1958, Xinhua banyuekan, Vol. 133, 1958, No. 11, pp. 14–17.

Editorial, *Make a Good Job of Formulating the Plan for 1958 to solve Correctly the Major Contradictions at Present Existing in the National Economy.* Jihua jingji 1957, No. 9, pp. 1–4.

[2] Editorial, *We Must Strive to Reduce Urban Use of Grain.* Renmin ribao, September, 2nd, 1957.

Editorial, *Harvest Work Must be Well Controlled.* Renmin ribao, September, 1957.

Supplementary Directive, *On Grain Unified Purchase and Supply,* October 13th, 1957. Xinhua banyuekan, Vol. 120, 1957, No. 22, p. 170.

[3] Ma Yin-ch'u op. cit., passim.

New China News Agency, *Discussion on the Question of Market Prices,* April 30th, 1957. Xinhua banyuekan, Vol. 108, 1957, No. 10, pp. 109–110.

being eroded very rapidly. Directives[1] had already called for less centralisation in planning the collectives, recommending that the brigade of 30–40 households (the former co-operative's size) should be the basic management unit, with considerable powers of decision making and of ownership. Draught animals were being returned to their former owners who were, once again, allowed to hire them to the collective. The higher investment rates, partly made possible by the abolition of such payments and, more important, rent payments, were further reduced as the collectives were forced to raise the share of consumption in production as an incentive measure. The private plots, larger in the socialist collective than in the semi-socialist co-operative, gave many peasants considerable independence. The situation called for drastic action to raise output and strengthen the collectives.

The Government closed the free market in August.[2] Directives[3] announcing an increase in grain levies argued that the presence of the private plot made it possible to do this without seriously reducing the standard of living of the peasants. Town rations were cut. At the same time, long term strategy towards the whole question of China's economic development was reconsidered. The State Planning Committee, which in September 1957 described the food situation as " still fundamentally tense ",[4] was clearly preoccupied with the rate of

[1] See three very important directives issued September 14th, 1957. Renmin ribao, September 16th, 1957:

(i) *On Rectification in the Agricultural Producers' Co-operatives.*

(ii) *On Making a Good Job of the Work of Production Management in Agricultural Producers' Co-operatives.*

(iii) *On Thoroughly Carrying Out a Policy of Mutual Benefit inside Agricultural Producers' Co-operatives.*

[2] State Council Decision *On Not Allowing Commodities, Subject to State Planned Purchase and Planned Supply, and Unified Purchase, along with other Commodities, to Enter the Free Market.* Issued August 9th, 1957. Released August 18th, 1957. Xinhua banyuekan, Vol. 116, 1957, No. 18, pp. 207–208.

See also Editorial, *The Free Market Must be Strictly Controlled*, Renmin ribao, August 18th, 1957.

[3] Supplementary Directive of October 13th, 1957, op. cit.

[4] Editorial, Jihua jingji 1957, No. 9, pp. 1–4, op. cit.

growth of agriculture in relation to the demands of the economy, given the aim of rapid industrialisation. The solution adopted by the Government was to mobilise labour in rural areas on a massive scale, concentrating on the crucial works of water conservation and fertiliser application.[1] A socialist education drive[2] in the countryside, involving an intensification of the class struggle, was the corollary. They were measures which led directly to the formation of the communes.

The short-lived " Hundred Flowers " period in 1957 was followed by the Anti-Rightist Campaign, which reached the villages during the Summer and Autumn of 1957. Rectification in the collectives and an intensification of the class struggle were combined with occasional calls for careful treatment of the highly-skilled middle peasants who still, according to one directive,[3] comprised 20% of the rural population. Autumn 1957, however, was more characterised by the attacks on the " rightist opportunist, capitalistic tendencies " of the middle and rich peasants during the so-called " two roads struggle " (a struggle for supremacy between capitalism and socialism in the countryside). The focus was on the pre-occupation of these peasants with the large private agricultural sector which had grown up under the 1957 policy of increasing incentives. During the latter part of 1957, Spring 1958 and especially

---

[1] Directive, *Decision on a Large Scale Agricultural Field Water Conservation and Fertiliser Collection Campaign This Winter and Next Spring*, September 24th, 1957. Xinhua banyuekan, Vol. 118, 1957, No. 20, pp. 149–150.

Also Jihua jingji, 1957, Nos. 7–12 passim, especially Hsiao Yu, *How to Allocate Agricultural Investment*. Jihua jingji, 1957, No. 9, pp. 5–8 and Wang Kuang-wei, *Views on Allocating the Agricultural Labour Force*, op. cit.

[2] Directive, *On Carrying Out Large Scale Socialist Education for the Entire Rural Population*, August 8th, 1957. Xinhua banyuekan, Vol. 115, 1957, No. 17, pp. 1–3.

[3] Directive, *On Thoroughly Carrying out a Policy of Mutual Interest inside Co-operatives*, September 14th, 1957. Xinhua banyuekan, Vol. 117, 1957, No. 19, pp. 137–138.

during Autumn 1958 (when the details of the abolition of private plots were published) the leadership decided to curtail the private sector. Evidence was now produced to show that private agricultural activity had grown so much as to constitute a threat to collective agriculture. Cases were quoted of peasants opposing the collectivisation of " excess " private plot; of collective agricultural income falling below private agricultural income.[1] The serious results from competition for labour, manure and water between the private and collective economies were stressed. A Hunan collective reported[2] that this competition caused a 10·8% decline in the rice yield compared with the unusually low level recorded in the drought of 1956. In Fukien,[3] 200 *mou* of paddy were killed by drought because labour and water were monopolised by the private plots. One brigade in a Fukien collective[4] ran a grain deficit in 1957 for the same reason, whereas in 1956 it had a surplus. Collective pig rearing was decreasing. Only 3% of the pigs in the Province of Kiangsu during Autumn 1957 were collectively owned.[5] The private sector was further attacked as a snare both to poor peasants and to party cadres, who still had bourgeois aspirations.

It is impossible to say whether the private plots were abolished all over China during the Great Leap Forward in 1958. The evidence certainly leads to the conclusion that abolition was widespread. Case studies of how communes were

---

[1] *Actively Develop the Collective Economy*. Renmin ribao, February 13th, 1958.

[2] Ibid.

[3] Amoy University Department of Economics, *The Complete Elimination of the Vestiges of Private Ownership of the Means of Production is a Requirement of the Development of Productive Power*, op. cit.

[4] Ibid.

[5] *Actively Develop the Collective Economy*, op. cit.

formed usually told a similar story.[1]  Activity in the Spring
water conservation drive was said to have made it " impossible
and unnecessary " to work the private plot and individually
rear pigs.  Similarly, high income from the public sector
reduced the profitability of working the private plot.  Poor
peasants " requested " the cadres to absorb the private plot
into the collective as a matter of urgency, such a policy being
opposed by the rich and upper middle peasants, again proving
their political unreliability.  The abolition, it was now
argued, no longer meant a drop in food consumption since the
commune messhalls, supplying free food, made adequate
provision.  The private plot was condemned as an inefficient
use of land, impeding the economic development of the collective sector.  Its elimination was approved as " rational ",
bringing productive relationships into line with production's
needs.  Both the model rules for communes[2] and the Party
resolution[3] of August 29th, 1958, recommended the com-

[1] A great deal of literature on this topic was published between August
and December, 1958.  See especially the following:—

Wang Yen-pu, *Survey Research on the Question of the Transition of Collectives
Towards a System of Public Ownership.* Xin jianshe September, 1958,
pp. 1–7.

Xin jianshe October 1958, passim.

Wu Chih-pu, *From Collective to Commune*, Hongqi, 1958, No. 8, pp. 5–11.

Niao Chia-p'ei et. al., *Attempted Discussion on the Revolution in the Rural
Distribution System during the Communisation Movement.* Jingji yanjiu, 1958,
No. 10, pp. 1–7.

Jingji yanjiu 1958, Nos. 11 and 12 passim.

Wu Pai-lin, *From Feng Ch'i Hsiang View the Inevitability of the Appearance
of the Communes.* Xuexu yuekan (Learning Monthly) 1958, No. 10, pp.
26–30.

Xuexu yuekan 1958, Nos. 11 and 12 passim.

Amoy University Department of Economics, *The Complete Elimination of
the Vestiges of Private Ownership* etc., op cit.

P'an Sheng-hwa, *Talking about the Change in People's Thoughts after the
Establishment of the Communes.* Xuexu luntan 1958, No. 4, pp. 30–32 and p.
10.  P'an states " The greatest difference between the collective and commune then, is that the commune has completely eliminated the private
economy".

[2] *Draft Regulations of the Weihsing Commune*, op. cit.

[3] Central Committee *Resolution on Questions Concerning the Establishment of
the Communes*, August 29th, 1958, op. cit.

munisation of the private plot, without compensation. Scattered trees might be retained by private owners. A People's Daily Editorial[1] of September 4th, however, commenting on the commune regulations, urged caution in absorbing the private plot, although it agreed that generally this should be done. The rules regarding livestock were more obscure. Livestock were to be communised except " small numbers of domestic animals." Pigs were not mentioned specifically, but with the abolition of the private plot it is clear that the commune was to become the major source of supply.

Again a retreat on the 1956–57 pattern quickly came in the form of the Central Committee's " Resolution concerning certain Questions in the People's Communes," of December 10th, 1958.[2] The theme of this resolution was that " reckless change " to communism, before the " objective existence " of the necessary conditions, was to cease. These conditions—a vast output, reduction in working time, industrialisation, mechanisation and electrification in agriculture, would not be fulfilled for many years. Some cadres were already prematurely implementing communist policies. For example, they were inflating the percentage of total income obtained from " free supply " (according to need) as opposed to that from work done. This was reducing the incentive to work. They were communising peasants' personal belongings, including clothes, quilts, household utensils and savings. These " crude " attitudes were condemned. The *collective* ownership system of the communes, with communist elements, was stressed. It was repeated that peasants were to be allowed to own the scattered trees round their houses and to rear small numbers of domestic animals. The private plot was not specifically mentioned. It had almost certainly been replaced by the messhall vegetable gardens which the resolution described. Messhalls were to rear their own pigs and poultry. Altogether the resolution called for a slower pace of change and of activity. The clause stipulating that peasants should be guaranteed

[1] *From Weihsing Commune Model Regulations see how to Run the Communes.* Renmin ribao, September 4th, 1958.

[2] Xinhua banyuekan, Vol. 146, 1958, No. 24, pp. 3–11.

eight hours sleep was a telling indication of the excesses of the campaign.

Communisation had been carried out largely during the month of September, when reports from the Provinces led the central authorities to believe that the Summer harvest had been a bumper one, unparalleled in China's history, while the Autumn harvest was going to be the same. The first plan for agriculture in 1958 had set targets of 196 million tons for food grains and 1·75 million tons for cotton. These were extraordinarily ambitious targets, in relation to the trend[1] in output actually achieved, according to Peking figures, 1953–57, of 2% per annum for food grains and 3·7% per annum for cotton. The 1958 plan for food grains called for a 5·9% rise on the 157 million tons of 1957 and a 6·7% increase for cotton. But in March 1958 when the Great Leap Forward, with its slogan "let politics take command," had become the dominant theme in policy, these targets were increased. That for food grains now became[2] 212 million tons, 15% above the 1957 level of output; the new cotton target was 1·97 million tons, 20% higher than the level attained in 1957. In January 1959 Liao Lu-yen announced[3] that actual output in 1958 was 375 million tons of food grains, representing a rise on 1957 of no less than 102%, and 3·35 million tons of cotton—or a 104% rise. Even more astonishing than these claims were the 1959 targets then proposed by Liao Lu-yen. The food grain production plan was for 525 million tons or a 40% increase on the level claimed for

[1] Perhaps the 5 year linear trend line by itself is not a fair measure of growth. The following, therefore, are official Peking figures of actual yearly output.

|  | Food Grains | | Cotton | |
|---|---|---|---|---|
|  | Output (million tons) | % Change on Previous Year | Output (million tons) | % Change on Previous Year |
| 1953 | 157 |  | 1·17 |  |
| 1954 | 160 | + 2·0 | 1·06 | − 9·4 |
| 1955 | 175 | + 9·5 | 1·52 | + 43·0 |
| 1956 | 182 | + 4·0 | 1·44 | − 5·2 |
| 1957 | 185 | + 1·7 | 1·64 | + 14·0 |

[2] Hsueh Mu-ch'iao, *How Does Statistical Work Make a Great Leap Forward?* Tongji gongzuo 1958, No. 5, pp. 1–5.

[3] Liao Lu-yen, *The Tasks of the Agricultural Front in 1959.* Hongqi 1959, No. 1, pp. 11–18.

1958. Cotton output planned for 1959 was 5 million tons—a 49% rise on the 1958 claim. These were to be attained on a *reduced* area of land compared with 1958 for grains and on the same area as 1958 for cotton. Labour productivity in agriculture was to be doubled in the year. Liao pronounced that these targets were not only " possible " but capable of overfulfilment. Targets, he stressed in the same speech, must be set with regard for " objective reality "; " pretentiousness " must be renounced.

In fact, it was already clear to one provincial Government at least[1] that the figures for the 1958 harvest were grossly inaccurate. The organisation of agriculture during the Autumn of 1958 had been so severely dislocated by the establishment of the communes and the native iron and steel furnaces[2] that it may be safely concluded that by no means all

[1] This was Kwangtung Province. On December 30th, 1958 the Kwangtung authorities claimed that the 1958 grain output was 34·5 million tons, (*The Successes of Kwangtung in Building up Agricultural Production during the Past Year*. Hong Kong Dagong bao December 30th. 1958.) On February 16th, 1959 at the Province's Communist Party Conference they amended this to 30·5 million tons, or 12% less than the previous figure. (*Grain Production This Year will Rise by 60% in Kwangtung*. Hong Kong Dagong bao February 16th, 1959.)

[2] In August 1958 the Central Committee of the Chinese Communist Party set a new 1958 target for steel output at 10·7 million tons (the 1957 output claimed was 5·35 million tons). This target was to be attained by the mass building of small, native blast furnaces throughout the countryside. By September 1st, 1958, over 50% of the pig iron produced in China was from such primitive furnaces, and this was to be raised even further during September. In October (when the communes were already established over most of China) the campaign turned from pig iron to steel smelting. Thousands of small furnaces were now speedily built to make steel. For example, after communes had been set up in Szechuan, within 13 days, 60,000 native furnaces were constructed, while 8 million people (out of an approximate total population of 73 million) were organised into brigades to mine iron ore and smelt iron and steel. Cadres who attacked the campaign on the grounds that it would seriously affect the large labour supply needed by agriculture during the busy Autumn season were attacked as rightists: they had failed to understand that agricultural development " demanded " such a policy to supply the necessary steel. It was not until near the end of the year that the campaign was called off. Major documents covering the campaign are in *Xinhua banyuekan* Vols. 141–146, 1958, Nos. 19–24 passim.

the Autumn crop was gathered. To raise incentives in the villages during the latter part of Winter and the Spring of 1959, a rectification campaign[1] was carried out in the communes, the aim of which was to implement the policies outlined in the December resolution. The major excesses, relating to centralisation in planning, the use of resources (including labour), distribution, together with great dissatisfaction over the management of the messhalls, were to be curbed. Cadres were instructed to define the boundaries between individual and collective agricultural activity. Evidence was produced during the campaign giving some indication of the peasants' response to the full communisation attempted in Autumn 1958. One article[2] recorded that peasants had destroyed tools rather than give them over to the commune. This destruction was only stopped by allowing their retention as private property. The land formerly worked as private plots had fallen out of use.[3] Without such plots peasants had no opportunity to use their leisure time productively so that labour as well as land was being wasted. The urgent question of how to increase the supply of fodder, pigs and hence more fertiliser inevitably brought the discussion back to the need for private plots. Editorials and articles began to preach the doctrine that " small freedoms " produce good results.

[1] Ch'en Cheng-jen, *On a Large Scale Spread a Mass Type Rectification Campaign*. Xinhua banyuekan, Vol. 148, 1959, No. 2, pp. 21–24.

[2] *Report by Hunan Provincial Committee Rural Work Department on Spreading the Rectification Campaign Throughout Five Communes*. Xinhua banyuekan, Vol. 152, 1959, No. 6, pp. 18–21.

[3] This either suggests that private plot land was not genuine arable land, but odd lots with no alternative uses, or that the new policy to plant a smaller area of grain crops in 1959 had already been reversed; that the need to use all land for food had been realised. The latter is the more likely explanation. For example, the Kwangtung Government, in 1957, had announced that one of the main reasons for the reduction in food grain production in the first half of 1957 was that too much rice land had been allocated to private plots as part of the 1957 policy to raise incomes. See an editorial, Nanfang ribao (Southern Daily, Canton) July 23rd, 1957, calling for a survey of the area of rice land which had been transferred to private plots. Also T'ao Chu, (Governor of Kwangtung) *Kwangtung Struggles to become a Province of 1,000 catty yields in 10 years' time*. Renmin ribao, December 14th, 1957.

In February 1959 a meeting of *sheng* cadres and members of the Central Committee of the Communist Party (including T'an Chen-lin, Liao Lu-yen and Liu Juei-lung) was held at Chengchow to discuss agricultural techniques. Apart from a short communique[1] on its discussions this meeting was shrouded in secrecy. Later it was revealed to have been a very important policy making session. A series of articles in the Chinese press discussed the implementation of its directives in various regions. The main result of this conference which concerns this study was an admission that the public sector could not supply enough pigs needed for soil fertility. The slogan adopted was " rear pigs both privately and publicly, with public rearing as the main source, private as the auxiliary."[2] The resurgence of private pig rearing was accompanied by appropriate incentives. The six adopted in Szechuan and Liaoning Provinces[3] were recommended to other Provinces. These were: (i) Pigs must be bought from peasants for cash. (ii) Peasants selling pigs should receive a certain amount of pork. (iii) Pig manure must be paid for according to quality. (iv) An amount of " fodder land " should be allocated to members, either according to the 5% rule of the collectives, or to the number of pigs kept. (v) Peasants were to be given time to work the plot and rear the pigs. (vi) Commune veterinary assistance and sows would be made available. From May to August 1959 this campaign was carried on intensively. The " anxiety of cadres " in restricting private pig rearing during 1958 was, characteristically, now described as " excessive ".

The period of unrealistic targets and claims ended in August 1959 with Chou En-lai's admission[4] that the 1958

---

[1] Renmin ribao, February 25th, 1959, p. 1.

[2] For example, *Simultaneously Rear Pigs Publicly and Privately; Even out the Advanced and Backward.* Renmin ribao, May 28th, 1959, p. 3.

Ma P'ing, *Yang T'an Special District's Experience in Rearing Pigs.* Renmin ribao, May 28th, 1959, p. 3.

[3] Editorial, *Rear Public and Private Pigs Simultaneously; Propagate Domestic Animals and Poultry.* Renmin ribao, May 20th, 1959.

[4] Chou En-lai, *Report on the Amendment of the Major Targets in the 1959 Plan and on the Campaign for Still Further Developing an Increase in Production and Economy,* August 26th, 1959. Xinhua banyuekan, Vol. 163, 1959, No. 17, pp. 20–24.

national output figures had wildly exaggerated actual achievements. The January figure of 375 million tons for grain output was amended to 250 million—a drop of 33·5% on the original claim but still a 25% increase over 1957 output; cotton fell to 2·1 million tons, that is by 37·5%, but still representing a 28% rise on 1957. Similar reductions were made for other products. The targets for 1959 received the same treatment: that for grain was scaled down by 47·5%, cotton by 53·8%. The Central Committee[1] called for a great economy drive in Autumn, especially in the use of food and raw materials. In October, Tan Chen-lin announced[2] that China had been hit by the worst natural disasters since 1949. Drought in Hupei was the worst for 70 years; Kwangtung had experienced Spring rain " not seen for many decades." Li Fu-ch'un[3] estimated that one third of China's arable area was affected. Nevertheless grain output achieved in 1959 was put at 270 million tons (8% above the 1958 amended figure); the 2·41 million tons of cotton claimed were 14·7% above 1958.

Since 1960 a statistical picture of China's economic performance has become even more difficult to portray than that in the earlier years. The statistical system's breakdown, culminating in Chou's statement of August 1959, brought almost a complete end to the publication of statistics other than plant or commune statistics. In addition, the export of the best Chinese journals (particularly the informative planning and statistical journals) ceased. The national newspapers continued to be exported. It is largely from these that the evidence for the years since 1960 has been drawn.

1960 was undoubtedly the worst crisis year encountered by the Chinese Communist Government thus far, regarding the difficulties of feeding the population and supplying the cotton and edible oil industries with the necessary raw materials. The

[1] *Resolution on a Campaign for Developing an Increase in Production and Economy.* Renmin ribao, August 27th, 1959.

[2] T'an Chen-lin, *The Construction of Large Scale, Modernised Agriculture Has Begun.* Renmin ribao, October 29th, 1959, p. 2.

[3] Li Fu-ch'un, *Struggle to Continue the Leap Forward in Socialist Construction.* Renmin ribao, October 27th, 1959, p. 2.

target[1] involving a 10% rise over 1959 output both for grains and cotton was set in January 1960, but targets and planning had ceased to be meaningful; the only worthwhile aim was maximum agricultural production. Demand for food was difficult to control, with a 20 million increase in the urban population 1958–60[2] inclusive, and 70 million[3] people engaged in a water conservation drive throughout Winter 1959 and Spring 1960. The fear of famine induced communes to plant food grains to the exclusion of industrial crops, despite price increases for soya beans (of 7·5%), groundnuts (12·08%) and sugar announced[4] in November 1959, along with tax induce-ments. Communes were directed[5] to increase the planted area of industrial crops and to reclaim and use every available plot of land. More potatoes and vegetables were to be planted,[6] the latter area by 20% on 1959, in accordance with their " greater significance " in 1960 than in other years. The high yields of potatoes and vegetables obtainable, that is their land-saving qualities, were stressed. Other land-saving innovations such as underplanting of trees and inter-row cultivation were adopted. One suggestion[7] that livestock rearing units should

[1] Li Fu-ch'un, *Report on the Draft Plan for 1960*, March 30th, 1960. Renmin ribao, March 31st, 1960, pp. 2–3.

[2] Editorial, *The Whole Party and Every Person is to be Occupied with Agri-culture and Grain.* Renmin ribao, August 25th, 1960, p. 1.

[3] Ibid.

[4] *Why Must Several Commodities' Prices be Amended?* Renmin ribao, November 16th, 1959, p. 3.

[5] Renmin ribao Editorials, *People's Communes Must Formulate Land Use Plans*, March 17th, 1960, pp. 1 and 3.
*Rationally Allocate Agricultural Production*, April 3rd, 1961, p. 1.
*Struggle to Overfulfil the Plan for Sown Area of Industrial Raw Material Crops*, April 21st, 1960, p. 1.

[6] For example, Editorial, *We Must Still Plant Somewhat More Potatoes.* Renmin ribao, May 4th, 1960, p. 3.
*Directive on the Autumn Production of Vegetables*, July 19th, 1960. Renmin ribao, July 20th, 1960, p.1.
Editorial, *Strive to Increase the Production of Vegetables in Autumn*, ibid., p. 2.

[7] Editorial, *Successively Reclaim and Plant Land; Strive to Attain an Abundant Harvest the Same Year.* Renmin ribao, May 10th, 1960, p. 7.

be transferred to the hilly districts to release land for cereals, seems to have overlooked the effect this would have on soil fertility.

The livestock situation was deteriorating, according to an important People's Daily Editorial of July 16th, 1960.[1] The pig death rate was so high in 1960 (over 50% among young pigs) that the total was declining in many places. The editorial openly conceded that poor management, giving rise to disease, was the main cause. Commune rearing units were admitted to be too large for good management, while over optimistic estimates had been made concerning the supply of fodder crops (some of which had fuel as a competing demand). Large quantities of animal fertiliser were being wasted because of the shortage of labour to collect and transport it to the fields.

After the Summer of 1960 an attempt was made[2] to channel all possible labour into agriculture, by cutting employment in industry and administration. The Province of Kwangtung alone was said[3] to have put back 1 million people into the fields during 1960. In Kiangsu[4] a commune released 50% of its industrial and service employees for agriculture. As well as this " push ", a " pull " effect operated—for redundancy was growing in industry because of retrenchment in construction and raw material shortages. The agricultural production front was strengthened by further decentralisation in the communes in November and December when the large production brigade (the equivalent of the former collective) began to return as the

[1] Editorial, *Grasp the Favourable Opportunity to Develop Pig Rearing.* Renmin ribao, July 16th, 1960, p. 1.

[2] Editorial, *Use the Majority of the Labour Force in the Fields.* Renmin ribao, July 27th, 1960, p. 1.

Many other articles are to be found in Renmin ribao during the period July-December, 1960. See especially August 22nd, 1960 and August 24th, 1960 for typical articles.

[3] Renmin ribao, July 27th, 1960, op. cit.

[4] Ibid.

basic planning, ownership and distribution unit.[1] Cadres from higher layers of commune administration were posted in the large production brigades, the production brigades and the yet smaller production teams.

[1] Yu I-fu, *To Fit in with New Circumstances We Must Have New Methods of Direction*. Renmin ribao, September 21st, 1960, p. 7. (Yu was First Secretary of the Kirin Party Committee).

Sun Min et. al., *Reform the Method of Direction by the Small Production Brigade*. Renmin ribao, November 10th, 1960.

Editorial, *Fully Develop the Combat Role of the Small Production Brigade*. Renmin ribao, November 25th, 1960, p. 1.

Editorial, *Use the Labour Force Rationally; Constantly Raise Labour Productivity*. Renmin ribao, December 18th, 1960, p. 1.

Editorial, *The Three Tier System of Ownership, Taking the Production Brigade as the Basis, is the Fundamental System in the Present Stage of the Communes*. Renmin ribao, December 21st, 1960, p. 1.

These titles of "production brigades" are confusing. This arises partly because of the different sizes of commune (some covered an entire *hsien*, others a *hsiang*) and consequently the number of tiers of management within the commune. It is clear from the context, however, that at this stage—the above articles were announcing that the "large production brigade" (see pages 18–19) was to be the basic unit of distribution and accounting. The fact that the Chinese includes the word "small" to describe the production brigade does not mean that it refers to the "production brigade" (former co-operative) as defined on page 18. This step took place during the latter half of 1961 and 1962.

# 6

## 1961 - 1962 :
## Decentralisation in the Communes
## The Restoration of the Private Plot—
## in Theory and Practice

NATURAL disasters in 1960 were claimed to be more serious than in 1959. They were described as the worst for 100 years.[1] Without producing any figures, Li Fu-ch'un conceded[2] in January 1961 that the 1960 targets for agriculture had not been achieved. He called upon the " entire nation " to strengthen the agricultural front to reverse the downward trend in crop production. Po I-po, Chairman of the State Economic Commission, presented[3] the industrial tasks for 1961. Industrial development was now to be suspended apart from raw materials (especially coal and metals), chemicals and transport. Construction was to be concentrated on a much narrower front. Central economic planning returned, after its dissolution

---

[1] *Report of the Central Committee of the Chinese Communist Party held January, 14–18th, 1961.* Hongqi, 1961, No. 3–4, pp. 1–3.

[2] Reported ibid.

[3] Po I-po, *Attain New Successes in China's Industrial Production and Construction.* Hongqi, 1961, No. 3–4, pp. 19–25. On the importance of adhering to the national plan Po I-po stated: " If any sector, any unit departs from the National Plan, from the demands of the whole (economy), unilaterally develops its own so-called " initiative ", then this is both contrary to the interests of the whole and must also be contrary to the interests of that particular sector itself." This was a complete reversal of the Great Leap Forward approach to local initiative.

during the Great Leap Forward. The use of resources was now to be strictly according to the plan. In agriculture the reorganization of the communes (even their disappearance, except in name) continued, in an attempt to raise local incentives, the efficiency of agricultural planning and production. Another rectification campaign, to " strengthen relations " between the cadres and masses, had already begun. Probably the most important incentive for the peasants, however, was the return of the private plot, together with the free market in the form of " rural fairs ".[1]

In China the number of articles published on a subject is a reliable indication of its importance. In 1961, applying this test, the private plot became a crucial issue in agriculture. The Economics section of the Academy of Sciences in 1962 referred[2] to this as " one of the important issues discussed by economists in 1961." The private plot's reintroduction was a sure indication of the low level of food production and incentives prevailing in the countryside. It represented a major retreat for the Government; an admission of the political and economic failure of the communes and a return to the 1957 level of socialism in agriculture at most. The retreat was accompanied by lengthy statements of doctrine, which restored the plot into Chinese Communist ideology. These wer efull of interest, illustrating the adaptability of Communist theory in the face of necessity. Was the private plot socialist in character, or had the " tail of capitalism " returned? What was its role and rationale in the socialist economy of China? These were discussed at great length. The conclusions reached must be listed briefly.

[1] Ch'en Hsing, *With Direction and Planning, Open Rural Fairs.* Renmin ribao, November 25th, 1960, p. 7.

Sung Lin, *Develop Rural Fairs; Enliven the Village Economy.* Renmin ribao January 18th, 1961, p. 7.

Kuan Ta-t'ung, *Problems Concerning Rural Household Subsidiary Industries, the Retained Plot and Rural Fairs.* Dagong bao, July 5th, 1961. This fits the rural fairs into Communist ideology—stressing the difference between these and capitalist markets.

[2] *Problems Concerning Commune Members' Household Subsidiary Industries and the Retained Plot: a Summary of the Views in Discussions among Economists during the Past Year.* Dagong bao, January 10th, 1962.

(i)  The nature of the private plot.

To some participants[1] in the debate the private plot was socialist in character: " one form of socialism " or " a special type of collective socialist ownership." The majority opinion,[2] however, was that it was an example of " individual " as opposed to " capitalist " ownership, but not entirely socialist; to speak of it as socialist was " not quite correct ". The plot was not capitalist because it was still owned by the commune and loaned for a long period to the peasants. It could not be rented or sold. No labour exploitation was involved. The produce of the plot was mainly consumed by the household, the surplus sold at " rational prices ". It was agreed that fears that the existence of the plot would lead to " spontaneous cap-italism " were ill-founded. The commune's control over the use of resources on the plot made this impossible. Furthermore, peasants now possessed a " definite sense of organisation, discipline and political ideology."[3] The private plot was but an example of the traces of the old society lingering in China during its transition from socialism to communism.

[1] For example, *On the Question of Members' Retained Plots in the Communes* Dagong bao, April 10th, 1961. Also Dagong bao, January 10th, 1962, op cit. This is a very good summary of the discussion.

[2] It is impossible to refer to all the articles on this subject. The following selection cover all the points raised.

*The Nature of Members' Retained Plots and Household Subsidiary Industries.* Dagong bao, March 15th, 1961.

Wang Te-pei, *Some Information on the Nature of the Retained Plot.* Dagong bao, May 19th, 1961.

Hu Shih-hao, *Completely Understand the Nature of the Retained Plot.* Dagong bao, May 26th, 1961.

Shih Hsiu-lin, *Some Views on the Nature of the Retained Plot.* Dagong bao, June 21st, 1961.

Kung Wen, *Talking About Village Household Subsidiary Industries* Part I. Gongren ribao, August 2nd, 1961; Part II, ibid., August 3rd, 1961.

Hsiao Liang, *Can the Development of Household Subsidiary Industries Help to Spread Spontaneous Capitalist Tendencies?* Gongren ribao, November 21st, 1961.

Li Hsuan et. al., *On the Prior Development of the Collective Economy, Actively Develop Household Subsidiary Industries.* Zhongguo nongbao, 1962, No. 2, pp. 28–30.

[3] Dagong bao, January 10th, 1962, op. cit.

(ii) The rationale and role of the private plot.

The main argument here is summed up in the following quotation from one of the contributors to the discussion: " In the present stage of the rural communes, to allow members to run small private plots is in accordance with the present objective economic conditions and is a common request of the vast mass of members."[1] The same arguments as those adduced (for example by the Academy of Sciences) in 1957 were repeated. The private plot satisfied demands which could not be met by the public sector in the present state of backwardness. For the plot to be the source of livestock products, vegetables and raw materials for household sideline occupations while the commune land concentrated on the production of food grains, was considered the most efficient division of labour. It was a valuable source of cash income for peasants. It utilised land and labour which might otherwise remain idle. Finally, it was a means of solving the livestock fodder shortage. Provided that the competition for labour and manure was reconciled its existence would actually promote the development of the collective economy by raising incentives.

The plot was to be distributed according to the 5% limit in the regulations for collectives. No reference was made to the 10% rule which was still in force when the plot was abolished in 1958. The sudden focus of attention on the plot at this time and repetition of the criterion to be used suggests that the instructions to cadres in the Spring of 1959 had not been widely implemented. Only a little evidence was found of private plots existing in 1959 and 1960. The 1959 decision to bring back the plot was certainly not put into effect immediately. Throughout 1960, articles called for the restoration of the plot or for plots of realistic size. Cadres were criticised for not carrying out instructions. The contrast between these references and those of 1961, however, is striking. In 1961 the urgency of the situation was unmistakable. The weaknesses of the collective economy, the need to raise incentives, and to grow food, recognition of the peasants' desire to own some land—were all openly acknowledged. The little evidence

[1] *On the Question of Rural Commune Members' Retained Plot.* Dagong bao, April 10th, 1961, op. cit.

available concerning the actual size of plots granted suggests that, compared with 1956 at least, large plots were allocated. Some examples were given in Table III, Chapter 2. In contrast to the earlier years, peasants were now allowed to grow any crops on the plot—from grain to economic crops,[1] such as tobacco. The introduction of the village fairs in Winter 1960–61 was designed to encourage peasants to sell some of the private plot produce and so " enliven " rural trade which had evidently declined under the communes.[2]

In Spring 1961 the formula " take publicly reared pigs as the main source, privately reared as the auxiliary " was completely reversed.[3] However successful the commune system had been in such matters as mobilising labour for water conservation projects and to fight natural disasters, it had not been successful in raising pigs on a large scale. The demand for skilled labour by large pig-rearing units was too great. The high costs and death rates of earlier years had persisted. With all efforts geared to food grain production, fodder supplies were not available from the commune land. Incentives to encourage private pig rearing were introduced[4] yet again, in the form of pork tickets for peasants selling pigs to the State and high prices for pig manure delivered to the collectively owned land. One report cited[5] the return to private hands of pigs com-

[1] Dagong bao, April 10th, 1961, op. cit. Contrast this with Teng Tzu-hui, *Concerning Extended Reproduction in Collectives and Several other Questions.* Xinhua banyuekan, Vol. 121, 1957, No. 23, pp. 155–159.

[2] For analysis of this decline in inter-commune trade see sources for Spring 1959 discussing the " law of value " and importance of " commodity production " under the commune system. A collection of such articles is found in a publication of the Chinese Academy of Sciences Economic Research Institute: *Collected Essays of Chinese Economists on the Question of Commodities, Value and Prices under a Socialist System.* Two volumes, Peking, 1959.

[3] *Again Talking about the Way to Rear Pigs.* Renmin ribao, June 14th, 1962, p. 2.

[4] See for example an Editorial Renmin ribao, February 24th, 1961, p. 1, where a report from Kwangtung is also published.

[5] *Public and Private Activity Together Raise a Great Many Pigs.* Dagong bao, March 24th, 1961, p. 1.
Wen Hwa, *Some views on the Present Rearing of Pigs.* Dagong bao, June 12th, 1962, p. 2.

munised in 1958 and 1959. The change was in line with the People's Daily Editorial[1] of January 6th, 1961 which called fertiliser the key to the 1961 harvests. Many articles stressed the importance of pigs as the main source of fertiliser, comparing a pig to a small chemical fertiliser factory.

The large production brigade (former collective) as the basic unit of ownership, planning and distribution[2] gave way to the smaller brigade, or former co-operative, while its subdivision, the production team, equivalent in size to the mutual aid team of the early 1950s, received more powers and influence over day to day agricultural work. The distribution system was based on the principle of " distribution according to labour," as in the collectives, but different from the 1958 commune system of free supply and time wages.    Following a great amount of job evaluation more emphasis was now placed on relating reward to work actually accomplished.  In principle this was the same as the collective method, but it was probably more efficiently administered in 1961 and 1962.  The most important aspect of this decentralisation in the communes was that it gave back to the small village units the powers over day to day management decisions which they had enjoyed in 1955 and 1956, but which were eroded in 1957 (as bigger and bigger collectives were formed) and finally in 1958. References were made[3] to the influence of " subjective beliefs " in the hitherto overcentralised system; to " man made reductions in output "

[1] *The Collection of Fertiliser is an Important Item of Work* etc., op. cit.

[2] For a discussion of this change and the consequent role of the large production brigade see

*Strengthen the Construction of the Production Brigade.* Nanfang ribao, March 14th, 1962.

*Correctly and Actively Make Use of the Authority of the Large Production Brigade.* Nanfang ribao, March 18th, 1962.

[3] Shu T'ai-hsin, *The Role and Powers of the Production Brigade in Production Management.*  Gongren ribao, July 26th, 1961.  Three other important articles are:

Editorial, *Respect the Powers of the Production Brigade.* Renmin ribao, June 21st, 1961, p. 1 (and a report from Kiangsi, ibid., pp. 1 and 3).

Ko Chih-ta, *Analysis of the Extension of Mass Type Economic Activity.* Dagong bao, October 25th, 1961, p. 3.

A report on a Hunan *hsien* headed *Work Research.* Renmin ribao, June 11th, 1962, p. 2.

resulting from the issue of unsuitable directives.  In future the crucial decisions regarding the use of land and agricultural technique were to be in the hands of those peasants who had expert knowledge of that area: those who knew the " natural laws " governing agriculture.  Secondly productivity of the small production unit was now to be clearly reflected in the living standard of that unit.  The trend to equal distribution of rural living standards between villages of widely different production levels was reversed.

Finally, in 1961 and 1962, the Government publicly adopted what had in fact been its planning sequence of the previous few years: agriculture, light industry, heavy industry. A hint[1] of this in August 1961 was followed[2] in December 1961 by a long statement reconciling the sequence with Marxist doctrine.  It argued that the reversal of priorities was quite consistent with the basic socialist economic law of planned proportional development which put heavy industry first. Agriculture in China had a " specially important status " which was to remain as a long term principle.  It was not a temporary expedient forced upon the Government by the natural disasters. Thus agriculture was finally given top priority in planning theory as well as in actual practice.  Policy had moved a long way since 1958.  It had moved in the right direction.

[1] Ou Yang-ch'eng, *On the Relationship between Production and Construction.* Dagong bao, August 21st, 1961, p. 3.

[2] Yang Chi-hsien, *The National Economic Plan Must Allocate According to the Sequence Agriculture, Light Industry, Heavy Industry.*  Dagong bao, December, 11th, 1961, p. 3.

## CONCLUDING REMARKS

The main conclusion to be drawn from the small body of evidence presented is that, in the current state of China's political and economic development, a private sector of agriculture, composed mainly of private plots, is a " necessary adjunct to the socialist economy."[1] There are two reasons for this. The first is that the peasants have shown themselves unwilling to surrender all private ownership of land, pre-sumably until they have enough confidence that the collective economy will supply their needs. This is not only a question of the output of the collective, but also of the Government's policy towards its distribution. The second reason is that the Govern-ment found that a private sector was needed to make up the deficiencies in the public sector. This was particularly true for fodder and pigs throughout the period under discussion, and for food during the crisis years of 1959–61 inclusive. The collective economy's excessive concentration on growing food grains, at the expense of industrial crops, had led to a shortage of fodder. This, together with poor management and high costs of large scale, collective pig-rearing enterprises, resulted in serious losses of pigs. A long term decline in the pig popu-lation and the corresponding reduction in fertilisers available, would have had a disastrous effect on crop yields. A division of responsibility was, therefore, needed, between the collective and private sectors, the former concentrating on the important food and industrial crops, while the latter produced vegetables and pigs.

As long as the private sector continued to exist, however, there was a danger that it would threaten the collective sector. The inexperienced collective administration was burdened with

[1] Editorial, *Under the Premise of Prior Development of the Collective Economy Develop Commune Members' Household Subsidiary Industries.* Renmin ribao, November 5th, 1961.

the task of controlling the competition between the two sectors for labour and fertiliser. The consequences of abolishing private activity, on the other hand, were also potentially serious. Alternative sources of fertiliser did not exist, so that collective pig farms would have to be immediately successful if supplies were to continue and to grow. There was the possibility that peasants, deprived of their plots, would sabotage the marketing schemes and slaughter livestock. Yet within two years, severe restriction of the private sector and even complete abolition was attempted.

The two cases were rather different. The 1956 attack on the private plot was contrary to the published rules of co-operatives and collectives, as well as to the declared aims of the leadership. When the attempted squeeze was later admitted, the Central Government claimed that it had not been responsible. The local cadres were blamed. In 1958, when the communes absorbed the plots, the situation was different. This was carried out according to the rules, with the open support of the Government. The case for elimination was that it had been made superfluous by the Great Leap; peasants had " requested " its abolition. But in the retreat of 1959 the local cadres were again criticised for under-valuing the contribution of the private plots as a source of vegetables and pigs. This criticism was repeated in 1961 when cadres were condemned for their dogmatic, subjectivist thoughts relating to private activities of this kind.

From the history of government policy outlined in part three it is possible to draw some general conclusions about the determinants and nature of policies carried out in China. The chronology of events described illustrate three important features of policy:

(i) The influence of theory.

(ii) The pragmatism of the Chinese leaders.

(iii) The weakness of the administrative system, which enabled the local cadres to implement policies different from those intended by the Central Government.

1. *The influence of theory over policy.*

The great influence of communist theory over Chinese agricultural policy can be seen by examining the events of 1955–62. The overriding belief was that only a *collective* agricultural economy could align productive relationships in such a way that the technical revolution needed in agriculture would be guaranteed. Collective ownership of the factors of production, together with *large* agricultural planning units, were fundamental prerequisites of growth, in the minds of the Chinese leaders. Thus, in announcing the establishment of the people's communes in 1958 the Party had only to state that their greatest feature was that they were " both large and publicly owned ",[1] for this was sufficient evidence of success. From the first establishment of the mutual aid teams in the early 1950s there was constant pressure on local cadres to merge teams, co-operatives and collectives into bigger planning units, together with more public ownership. These policies were carried out in the face of a formidable body of evidence that they would seriously affect production. Similarly, for ideological reasons policy was to ensure that the inevitable bourgeois aspirations of the peasants did not manifest themselves in " spontaneous capitalism ". Again though the evidence regarding the implications of such a policy was clear, excessive limitations were placed on the private sector of agriculture.

The vacillations in policy, the advances towards socialism and withdrawals, could be used to support the view that Chinese policies are, as the leaders assert, guided by dialectic theory. Progress, they argue, is obtained by moving through the sequence balance-imbalance-balance, and so on. The weaknesses in the economy and in society are exposed by the advances, after which they must be remedied before proceeding further ahead. This strategy involves moving to an extreme position, retreating when necessary and coming to rest temporarily at a position still in advance of the original point. It would be a mistake to dismiss this as a factor in Chinese policy. Most of the difficulties of rapid socialisation were accurately foreseen by the Chinese leaders. It is not inconceivable there-

---

[1] The Chinese is romanised " Yi da er gong."

fore, that just as the advances were planned, so were the steps backward. Perhaps it was considered that this strategy would condition the peasants to accept change. For example, the rapid change from co-operative to collective was accompanied by an encroachment upon private plots. Their restoration, even though this was merely in accordance with the rules could then be claimed as a " concession ". The collectives might be more acceptable to the peasants; the co-operatives, however, might now be completely acceptable. If the move from mutual aid team to collective was followed by a return to the co-operative— the advance would still represent a very real achievement.

### 2.   *The pragmatism of the Government.*

Granted the above theoretical basis for policy, the history of agricultural policy in Communist China also illustrates the pragmatism of the Government. Thus, the first attack on the private plot can either be explained as a miscalculation or as a failure of the Central Government to control local cadres. Assuming that secret orders were given to cadres to test the resistance of the peasants to complete land collectivisation, having discovered that such a policy reduced incentives, the Government wasted no time in reversing it. The events of 1958 may be viewed as a logical outcome of the economic and political situation existing in 1957, rather than as merely reflecting ideology. There were practical reasons for the drastic action taken, resulting in the Great Leap Forward, formation of the communes and abolition of the private plots. In chapter 5 it was shown that the relatively low rate of growth in agriculture during 1956 and 1957 had begun to raise doubts in the minds of the Government concerning the possibility of maintaining a rapid rate of industrialisation. In addition, the collectives showed signs of reverting to co-operatives. The growing population problem had to be incorporated into a strategy for growth. It could be argued, then, that the Government had little freedom of manoeuvre in 1958. The commune idea had a practical basis, even if it was not successful in most of its aims. Another example of the Government's pragmatism is its policy towards pig-rearing. The communes failed to rear pigs at costs as low as those of private households and in sufficient numbers. They also failed to grow enough fodder crops. In admitting these, together with the statement that the peasants' ideological level was not yet appropriate for full socialisation, the Government restored the private sector as the main source of supply of pigs—quite contrary to theory, but nevertheless realistic.

### 3. *The weakness of the administrative system.*

The third factor was the degree of control exercised by the Central Government over the lower layers, especially the local cadres. The troubles in collectives during the Spring of 1956 reflected the over centralisation in target setting. Cadres in collectives were presented by the *hsien* Governments with unattainable targets and had little power to revise them. The greatest index of a cadre's success was the fulfilment (or over-fulfilment) of the output targets. Though the rules for collectives stipulated that private plots should be allocated, cadres might have considered that these would impede the fulfilment of the all-important targets. It was worth disregarding one directive to ensure that another was carried out. The 1958 excesses in the communes can be attributed to the failure of the Central authorities to keep the local party cadres' zeal in check. They also reflected the atmosphere of leaping over stages on the road to communism, and of " politics in command " producing miraculous results. The instructions to cadres regarding the private plots, in any case, were vague. The regulations for communes stated that " generally " they should be communised forthwith. Other government statements stressed that immediate action was to be avoided if it was considered that output might be adversely affected. Thus local cadres had considerable powers of deciding what to do. In the context of anti-rightist and rectification campaigns, great leaps and the target economy, they were likely to err on the side of too much rather than too little socialism. If they allocated tiny private plots, or none at all, then at least they had no fear of being charged with encouraging the capitalist inclinations of the peasants. The cadres had to weigh the possibilities of succeeding in carrying out " advanced " policies, which would earn them praise from the higher echelons of the Party, against the chances of being criticised (and even dismissed) as conservative, if they adopted more cautious and yet safer policies. When local cadres were blamed for such problems as the decline in pigs, or the ill effects of excessively limiting private agricultural activity, their confidence must have been destroyed. The fluctuations in government policy towards the private sector were enough to make cadres reluctant to act at all, until it became clear that counter orders were not about to be issued. Government statements calling on local cadres to criticise their superiors, others exonerating them from their mistakes in view of their inexperience, and in general warning peasants that criticism of local cadres must not go too far, all provide evidence that demoralisation had reached a serious level during the early 1960s.

Since 1961 the Government's policy has been to give agriculture top priority in the economy. All other activities have been planned to serve its interests. The organisation of agriculture has focussed upon the small, basic level units of management—the production brigade and team. A private sector has been allowed to exist along side the collective sector. The statistical void makes it impossible to judge whether these policies are already producing good results. There is evidence that the supreme importance of chemical fertilisers has been fully recognised and that efforts are being made to increase supplies rapidly. If these are successful they will at the same time raise the level of agricultural output in the public sector (thus making it more acceptable to the peasants) and reduce the Government's need for the animal fertiliser of the private sector. When that point is eventually reached, the significance of the private sector will begin to disappear.

# *Appendices*

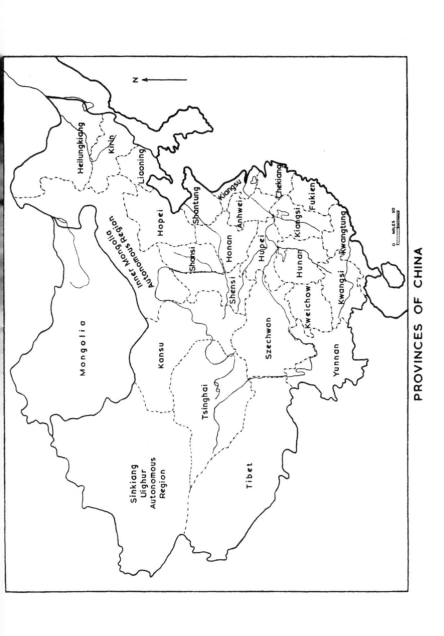

PROVINCES OF CHINA

# APPENDIX II

## LIST OF SOURCES CONSULTED

It did not seem useful to compile a bibliography of all the newspaper and journal articles used in preparing this monograph, especially as they have been referred to as footnotes to the text. The following is a list of the names of Chinese newspapers, journals and books used, while the few English language books and articles consulted are given in full.

## I. CHINESE LANGUAGE MATERIALS

### 1. NEWSPAPERS

Dagong bao: Impartial Daily Peking.
Dagong bao: Impartial Daily, Hong Kong.
Gongren ribao: Workers' Daily, Peking.
Nanfang ribao: Southern Daily, Canton.
Renmin ribao: People's Daily, Peking.

### 2. JOURNALS

Hongqi: Red Flag, Peking.
Jihua jingji: Planned Economy, Peking.
Jihua yu tongji: Planning and Statistics. This was a merger of Jihua jingji and Tongji gongzuo beginning 1959.
Jingji yanjiu: Economic Research, Peking.
Liangshi gongzuo: Grain Work, Peking.
Zhongguo nongbao: Chinese Journal of Agriculture, Peking.
Zhongguo nongyeh kexue: Chinese Agricultural Science, Peking.

Tongji gongzuo: Statistical Work, Peking.
Turang xuebao: Journal of Soil Science, Peking.
Xin jianshe: New Construction, Peking.
Xinhua yuebao: New China Monthly, Peking.
Xinhua banyuekan: New China Semi-Monthly, Peking.
  This succeeded Xinhua yuebao in 1956.
Xuexu luntan: Forum of Learning, Amoy.
Xuexu yuekan: Learning Monthly, Amoy.


3. Books

Agricultural Work Department: *Rural Work Questions,* 1955,
  Peking 1955.
Chiang Hsueh-mo: *The Distribution System of Socialism,*
  Shanghai, 1962.
Chinese Academy of Sciences Economics Research Institute:
  *Compendium of Materials on Agricultural Producers' Co-operatives
  during the Reconstruction Period of the National Economy,* 1949–52.
  Peking, 1957.
Chinese Academy of Sciences Economics Research Institute:
  *Selected Essays of Chinese Economists on the Question of Com-
  modities, Value and Prices under a Socialist System.* 2 volumes.
  Peking, 1959.
Chinese Committee for the Promotion of International Trade:
  *New China's Economic Achievements of the Past Three Years,*
  Peking, 1954.
*Collection of Selected Laws of the Chinese People's Republic.* Peking,
  1957.
*First Five Year Plan for the Development of the National Economy,
  1953–57.* Peking, 1955.
General Office of the Central Committee of the Chinese Com-
  munist Party: *The Socialist High Tide in The Chinese Country-
  side.* 3 volumes. Peking, 1956.
Hsu Ti-hsin: *An Analysis of the National Economy of China during
  the Transition Period.* Peking, 1957.
Ma Yin-ch'u: *My Economy Theory, Philosophical Thoughts and
  Political Standpoint.* Peking, 1958.
*People's Handbook for 1961.* Peking, 1962.
*Plan for Agricultural Development 1956–67.* Peking, 1960.

State Statistical Bureau: *Statistical Materials on Agricultural Co-operativisation and the Distribution of the Product in Co-operatives during 1955.* Peking, 1957.
State Statistical Bureau: *Ten Great Years.* Peking, 1959.


## II. ENGLISH LANGUAGE MATERIALS

1. BOOKS

J. L. Buck, *Land Utilisation in China.* Shanghai, 1937.
Ed. A. Datta, *Paths to Economic Growth.* London, 1962.
A. Eckstein, *The National Income of Communist China.* New York, 1962.
N. Jasny, *The Socialised Agriculture of the U.S.S.R.* Stanford, 1949.
N. Jasny, *Soviet Industrialisation 1928–52.* Chicago, 1961.
A. M. Khusro and A. N. Agarwal, *The Problem of Co-operative Farming in India.* London, 1961.
T. H. Shen, *Agricultural Resources of China.* New York, 1951.
Ed. E. F. Szczepanik, *University of Hong Kong Proceedings of the Symposium on Economic and Social Problems of the Far East.* London, 1962.
United Nations Food and Agriculture Organisation, *Annual Yearbook of Production,* 1959. Rome, 1960.

2. ARTICLES

P. Allan, *Fertilisers and Food in Asia and the Far East.* Span, Vol. 4, No. 1, 1961. London, 1961.
T. Balogh, *Agricultural and Economic Development: Linked Public Works.* Oxford Economic Papers, 1961, No. 1.
V. M. Dandekar, *Economic Theory and Agrarian Reform.* Oxford Economic Papers, 1962, No. 1.
A. Eckstein, *The Strategy of Economic Development in Communist China.* American Economic Review. May, 1961, No. 2.
N. Georgescu-Roegen, *Economic Theory and Agrarian Economics.* Oxford Economic Papers, 1960, No. 1.

J. A. Newth, *Soviet Agriculture: the Private Sector.*
  Part I, Soviet Studies, October 1961, No. 2.
  Part II, ibid., April 1962, No. 4.
N. W. Pirie, *Future Sources of Food Supply: Scientific Problems.*
  Journal of the Royal Statistical Society, Series A (General),
  Vol. 125, part 3, 1962.
H. L. Richardson, *The Use of Fertilisers in the Far East.* Pro-
  ceedings of the Fertiliser Society. No. 41. London, 1956.
H. L. Richardson in *Transactions of the International Society of Soil
  Science, Vol. 1, July, 1952. Joint Meeting on Soil Fertility.*
L. Volin, *Soviet Agriculture under Khrushchev.* American Economic
  Review, May 1959, No. 2.

# APPENDIX III

## CONVERSION RATES FOR CHINESE MEASURES

1. AREA    1 mou    = 10 fen.
               1 fen    = 10 li.
               1 li     = 10 hao.

               1 mou    = 0·067 hectares or 670 sq. metres.
               1 mou    = 0·1647 acres or 797 square yards.

               1 hectare    = 14·92 mou.
               1 acre    = 6·07 mou.

2. WEIGHT    1 tan    = 100 chin.
                 1 chin    = 0·501 kilogram.
                 1 chin    = 1·1023 pounds.

                 1 kilogram = 1·996 chin.
                 1 pound    = 0·989 chin.

# INDEX

Milton Keynes UK
Ingram Content Group UK Ltd.
UKHW042324061024
449327UK00004B/29